FAO ANIMAL PRODUCTION AND HEALTH / **MANUAL 27**

Manual for the management of operations during an animal health emergency

Authors
Lionel A. M. Gbaguidi
Susanne Münstermann
Modou Sow

Food and Agriculture Organization of the United Nations
Rome, 2022

Recommended Citation
Gbaguidi, L.A.M., Münstermann, S., Sow, M. 2022. *Manual for the management of operations during an animal health emergency.* FAO Animal Production and Health Manual No. 27. Rome, FAO.
https://doi.org/10.4060/cc0068en

ISSN 1810-1119 [Print]
ISSN 2070-2493 [Online]

ISBN 978-92-5-136207-5

Contents

Figures

Tables

Abbreviations and acronyms

AAR	After-action review
AH-EOC	Animal Health Emergency Operations Centre
CFSPH	The Center for Food Security and Public Health
CVO	Chief Veterinary Officer
DVS	Director of Veterinary Services
EMC-AH	Emergency Management Centre for Animal Health (FAO)
EOC	Emergency Operations Centre
FAD PreP	United States Foreign Animal Disease Preparedness and Response Plan (USA)
FAO	Food and Agriculture Organization of the United Nations
FEMA	Federal Emergency Management Agency
FMD	Foot-and-Mouth disease
GEMP	Good Emergency Management Practice: The Essentials
GIS	Graphical information system
HPAI	Highly Pathogenic Avian Influenza
IAP	Incident action plan
IASC	Inter Agency Standing Committee
ICS	Incident Command System
ICT	Information and communications technology
LEGS	Livestock Emergency Guidelines and Standards
MBN	Meat Board of Namibia
MoU	Memorandum of understanding
NAHEMS	National Animal Health Emergency Management System
NIMS	National Incident Management System (USA)
OFA	Office of WFS Follow-up and Alliances
PHEOC	Public Health Emergency Operations Centre
PIO	Public Information Officer
PPE	Personal protective equipment
PPEP	Progressive pathway for emergency preparedness
PVS	Performance of Veterinary Services
SitRep	Situation Report
SMEACS-Q	Situation, Mission, Execution, Administration/Logistics, Control & Communications, Safety - Questions
SOP	Standard operating procedure
TAD	Transboundary Animal Diseases
USDA	United States Department of Agriculture
WAHIS	World Animal Health Information System
WHO	Word Health Organization
WOAH	World Organisation for Animal Health (WOAH, founded as OIE).

For the purpose of this manual, definitions provided in the GEMP manual and the glossary of WOAH's Terrestrial Animal Health Code are relevant. Where necessary some additional definitions are included to ensure consistency and facilitate understanding of some terms.

Acknowledgements

The authors acknowledge with thanks the following expert advisers: Edgardo Arza, Amadou A. Ndiaye, John A. Ohemeng, Galib Abdulaliyev, Merab Acham, Lotfi Allal, Samantha Allen, Malcolm Andersen, Edem Apedwin, Alfonso Araujo, Hugo Araya Veliz, Lasha Avaliani, Charles Bebay, Guillaume Belot, Francesco Berlingieri, Jaouad Berrada, Laouad Berrada, Etienne Bonbon, Federica Borrelli, Abdoulaye Bousso, Marta C. R. Figueroa, Tony Callan, Maria Campuzano, Paul Cox, Ian Dacre, Paolo Dalla Villa, Vittoria Di Stefano, Seynabou Diack, Daniel Donachie, Esther Dsani, Dee Ellis, Danso Fenteng, Jean-Marc Feussom, Assane G. Fall, Nadav Galon, Andres Gonzalez Serrano, Jonas Gutschke, Keith Hamilton, Debbie Hill, Francisco Javier Reviriego Gordejo, Christine Jost, James K. Wabacha, Jessica Kayamori Lopes, Gael Lamielle, Eibhlinn Lynam, Naftaly M. Mwaniki, Ruben M. Zuniga, Jason Males, Arduino Mangoni, Hoang Manh Tien, Jean Marc Mfeussom, Rosanne Marchesich, Jered Markoff, Lijin Ming, Fred Monje, George Mukora, Lee Myers, Beatrice Nannozi, Cassimir Ndongo, Marius Niaga, Serge Nzietchueng, Pawin Padungtod, Pornpitak Panlar, Ago Partel, Rose Penda, Marie Pierre Doguy, Ludovic Plée, Barbara Porter-Spalding, Frédéric Poudevigne, Mariano Ramos, SéverineRautureau, Diego Rojas, Eric Rojas Torres, Maria Romano, Orr Rozov, Peter Rzeszotarski, Apanun Saeliu, Onpawee Sagarasaeranee, Ismaila Seck, Amy Snow, Batsaikhan Sodnom, Frida Sparaciari, Marcel Spierenburg, Hayley Squance, Nir Tenenbaum, Nguyen Thi Thuy Man, Paolo Tizzani, Jose Urdaz, Sophie VonDobschuetz, Wang Youming.

Special thanks to the information management and communication team involved in the editorial production: Claudia Ciarlantini and Cecilia Murguia.

Chapter 1
Overview

1.1 INTRODUCTION

This manual provides guidelines for countries and relevant local, national, regional and international organizations to prepare for and manage operations during animal health emergencies.

This manual is a compendium to the Food and Agriculture Organization of the United Nations (FAO) *Good Emergency Management Practice: The Essentials* (the GEMP manual) and provides an overarching view on how to address the peace time and emergency phases of an animal health event causing an emergency.

In addition to the GEMP manual, this manual is aligned to the *Livestock Emergency Guidelines and Standards* (LEGS). LEGS is a set of international guidelines and standards created to support the design, implementation and evaluation of livestock interventions during emergencies to help people affected by humanitarian crises.

The GEMP manual provides the basis for the FAO *progressive pathway for emergency preparedness* (PPEP).

The PPEP is a tool that enables national veterinary services to self-assess and standardize their animal health emergency management capacity needs and to achieve a sustainable state of emergency preparedness in line with the World Organisation for Animal Health (WOAH, founded as OIE) international standards (i.e. Terrestrial and Aquatic Animal Health Codes, the Manual of Diagnostic Tests and Vaccines for Terrestrial Animals and the Manual of Diagnostic Tests for Aquatic Animals).

Furthermore, the WOAH Performance of Veterinary Services (PVS) tool includes three critical competencies that are directly related to the management of operations during an animal health emergency:

1) Emergency funding (Chapter I – Critical Competency 9);
2) Emergency preparedness and response (Chapter II – Critical Competency 5); and
3) Internal coordination (chain of command) (Chapter I Competency 6. A.).

Not intended to be prescriptive, this manual supports veterinary services and competent authorities to achieve the standards articulated in the documents mentioned above, by providing examples for structures and workflows from good practices implemented by Australia, Canada, the United States of America, and a few other countries. In doing so, veterinary services and competent authorities can compare their own systems or use these examples as guidance to create or adapt their own systems.

EXAMPLE
Application of this manual

The Republic of Cameroon has successfully used this manual as a template to develop a national manual adapted to the Cameroon context.[†]

This manual has also served as a basis for simulation exercises in several countries (Australia, Cameroon, Chile, Colombia, Ghana, Kenya, Senegal, Thailand, Uganda, the United States of America and Viet Nam) and feedback from those exercises has been incorporated into this publication.

_____

[†] Manuel pour la gestion des operations lors d'une urgence zoosanitaire au Cameroun

This manual does not address every situation but aims at providing sufficient guidance for countries to customize their approach in preparing for and managing operations during an animal health emergency.

- **Chapter 1** is an introduction to this manual, identifying how it complements other animal health related documents, with a particular emphasis on the GEMP manual.
- **Chapter 2** describes the steps to be taken during peacetime to prepare for the management of operations during an animal health emergency and describes specific preparedness activities that are necessary for efficient management of operations during an animal health emergency.
- **Chapter 3** describes animal health emergency operations and how to implement mechanisms and systems to manage those operations during an animal health emergency.
- **Annex 1** presents a collection of checklists, forms and templates that could be referenced to help support the management of operations during an animal health emergency.
- **Annex 2** provides guidance for establishing an emergency operations centre (EOC).

1.2 SCOPE OF THE MANUAL

This manual focuses on preparing for and managing operations during an animal health emergency that results from animal diseases, infections or infestations, including zoonotic animal diseases.

This manual provides guidance on the 'preparedness', and 'response' actions undertaken through 'peacetime' and 'emergency' phases, which are described in the GEMP manual and summarized below.

Although broadly applicable to all hazards, this manual does not specifically address animal health emergencies as a consequence of natural disasters such as flood or drought, or industrial disasters such as nuclear accidents or chemical pollution, feed contamination with animal or public health impact, or food safety emergencies with animal implications. It also does not address scientific or technical aspects of an animal health emergency.

1.3 GOOD EMERGENCY MANAGEMENT PRACTICE: THE ESSENTIALS

The FAO GEMP manual sets out in a systematic way the elements required to achieve an appropriate level of preparedness and proposes an approach to animal health emergency management inclusive of all types of events, whether they are caused by natural phenomena, including non-infectious events, or by accidental or deliberate human action (FAO, 2011).

Definitions that are used in the GEMP manual and applied consistently in this manual are:

- **Animal health emergency** refers to a state where a major disruption or condition is triggered by an animal health event that can often be anticipated or prepared for, but seldom exactly foreseen. It may be caused by several hazards, including outbreaks of high impact transboundary animal diseases, both terrestrial and aquatic.
- **Emergency management** refers to the holistic organization and management of responsibilities, resources and actions that address all aspects of an emergency. Emergency management involves plans and institutional arrangements to engage and guide the efforts of public and private sectors in a systematic, comprehensive and coordinated way.

The GEMP manual provides good emergency management practices throughout all four phases of an animal health event causing an emergency. These phases are identified and described in the GEMP manual as: 'peacetime', 'alert', 'emergency' and 'reconstruction'.

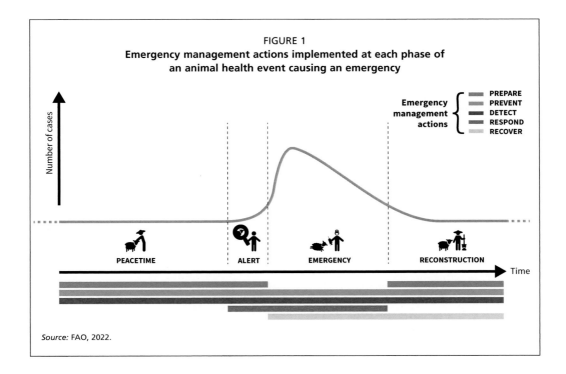

FIGURE 1
Emergency management actions implemented at each phase of an animal health event causing an emergency

Source: FAO, 2022.

These phases are illustrated in Figure 1, along with the emergency management actions that should be implemented at each phase. These phases and actions are briefly described below and explained in detail throughout the GEMP manual.

Peacetime phase refers to the period of time prior to a specific animal health event when no extraordinary or emergency actions are necessary for that event. It means, for example, that there may be peacetime on the front of one disease, while there is an emergency on the front of one or more other diseases.

Alert phase refers to the period of time when the level of risk due to an animal health event requires close observation of all activities, rapid transmission, sharing and assessment of relevant information and quick precautionary action to address an impending emergency. The alert phase is the period when a threat is advancing or has been identified. It can be triggered by suspicion of a case of a priority disease, or by a confirmed outbreak in the vicinity of the country or in trading partner countries. During the alert phase, an early warning system is used.

Emergency phase refers to the period of time requiring immediate actions to avoid or mitigate direct and indirect losses caused by an animal health event. Although only this phase is referred to as 'emergency', the entire event is addressed by emergency management.

Reconstruction phase refers to the period following the emergency phase and is dedicated to the re-establishment of animal populations, the recovery of pre-emergency (human and animal) health levels (including efforts to reduce risk factors), the relaunching of animal production systems, value chains and trade, the restoration of livelihoods, and the support to other socio-economic aspects impacted by the animal health event. The animal health status recovered may be different from that before the event. The reconstruction phase is also used to implement a post-experience assessment or an after-action review (AAR).

The GEMP manual explains that all through the animal health event, five types of emergency management actions should be implemented: 'prepare', 'prevent', 'detect', 'respond' and 'recover'. These actions are implemented in accordance with their relevance to each phase of the animal health event causing the emergency.

The **'prepare'** action refers to the development and implementation of strategies, policies, programmes, systems and analyses prior to an animal health emergency, in order to prevent, detect, respond to, and recover from said emergency. Prepare is the predominant action of emergency management during peacetime but it is also important during reconstruction.

The **'prevent'** action refers to the implementation of activities, programmes, and systems that enable an organization to avoid, preclude or limit (mitigate) the impact of an animal health event. Prevent is an action of emergency management that is key during the alert phase to avoid the event becoming an emergency, but it also applies as a general measure during the other phases.

The **'detect'** action refers to the implementation of activities, programmes, and systems to identify an incursion, emergence or re-emergence, or spread of a hazard, or define the level of presence or demonstrate the absence of the hazard. Detect is particularly important during the alert and emergency phase to know where to respond. It is also important in peacetime in terms of readiness and in the reconstruction phase to help recover a favourable animal health status.

The '**respond**' action refers to the implementation of activities, programmes, and systems aimed at the rapid containment and eventual elimination of the cause of an animal health event, and at the mitigation of its negative consequences. Respond is the primary action of the emergency phase. However, it is also possible to respond pre-emptively during the alert phase.

The **'recover'** action refers to the implementation of activities, programmes, and systems to relaunch animal production systems, value chains and trade, to restore livelihoods, and to support other impacted socio-economic aspects. Recover is predominant during the reconstruction phase but can be activated before the end of the emergency phase, especially if that phase is long-lasting or concerns large parts of the territory.

Chapter 2
Preparing to manage operations during an animal health emergency

2.1 INTRODUCTION
The ability to properly manage operations during an animal health emergency relies on the level of advance preparedness achieved. Veterinary services must invest in preparing their response capabilities at national, sub-national and local levels to manage operations effectively and efficiently during an animal health emergency.

This section addresses:

- prerequisites for timely and effective response;
- establishing response capabilities, which include:
  - authorities and enabling frameworks;
  - emergency response systems;
  - emergency operations centres;
  - emergency operations plans and supporting procedures;
  - systems to support operations;
- maintaining response capability, which include:
  - training of personnel;
  - testing through simulation exercises;
  - monitoring and review.

2.2 PREREQUISITES FOR TIMELY AND EFFECTIVE RESPONSE
Four basic prerequisites are needed to respond to an animal health emergency quickly and effectively. These include:

- identify, categorize and prioritize animal health events;
- establish triggers for declaration of an animal health emergency;
- early detection of an animal health event; and
- pre-determine response goals and strategies.

These prerequisites allow for early detection of an animal health emergency and enable officials to implement an appropriate and timely response.

2.2.1 Identify, categorize and prioritize animal diseases
The identification, categorization and prioritization of animal health events and their corresponding consequences, allow countries to determine the animal health events that

have the potential to cause an animal health emergency and justify the development and implementation of specific emergency preparedness and response plans.

Strategic risk assessments of animal health events should be undertaken as a prepared-ness action, whereby the outcome guides actions to prevent, prepare for and reduce the level of risk associated with a particular animal health event. The strategic risk assessment should include the potential consequences, such as socio-economic, food security and public health impacts. Further guidance on risk analysis in animal health emergency man-agement is included in Annex IV of the GEMP manual.

Developing a list of priority animal health events of greatest concern to the country including those that are zoonotic in nature is crucial to properly planning for operations during an animal health emergency. Once created, the priority animal health events list should be reassessed on a regular basis according to the most current strategic risk assess-ment.

2.2.2 Establish triggers for declaration of an animal health emergency

To trigger a declaration of an animal health emergency, an animal health event needs to meet a statutory definition.

Each country should identify the animal health events that are likely to lead to an animal health emergency and develop the adequate infrastructure (emergency organizations/units, operational strategies and response plans) needed to respond to and recover from animal health emergencies.

Countries should also define a scale or severity level for animal health emergencies. Defining the severity of animal health emergencies allows officials to implement, as soon as reasonably practicable, the appropriate level of coordination, resources and support necessary for the management of operations during an animal health emergency.

EXAMPLE
Severity levels for an animal health emergency

The 2018 Inter-agency Standing Committee standard operating procedures (IASC, 2018) (SOPs) use scale of urgency, complexity, capacity and risk of failure to define the levels and to deliver effectively to affected populations.

2.2.3 Early detection of an animal health event

The success of a country's capability to quickly and accurately recognize and characterize an animal health emergency depends on the following factors:

- a national surveillance programme for priority animal health emergencies, which is capable of detecting emerging and re-emerging animal health events that trigger disease investigations and laboratory diagnostic follow up;
- awareness of the country's priority list of notifiable animal health events by relevant stakeholders;

- completion of a risk assessment, including pathway analysis and mapping to identify hot spots for targeted surveillance (risk-based surveillance);
- availability of an early warning system to alert authorities and stakeholders;
- availability of laboratory diagnostic capacities, which should include an accreditation system for private laboratories (i.e. compliance with ISO 9000 and ISO/IEC 17025 General Requirements for the Competence of Calibration and Testing Laboratories; proficiency testing for the evaluation of a laboratory's capability to conduct specific diagnostic tests) and a formal surge support collaboration framework (e.g. WOAH Reference Laboratory Network, or FAO Reference Centres);
- trained staff capable of recognizing the importance and urgency of an animal health event, that may cause an animal health emergency; and
- the technical ability to perform epidemiological investigations and analyses, including the development of case definitions.

2.2.4 Pre-determine response goals and strategies

The animal health emergency response goals and strategies should be developed and documented for each of the animal health events previously identified and prioritized as part of the strategic risk assessment. These documents should provide information about the nature of the animal health event, the principles of its control, the objectives of the control policies and the impact these may have on animal welfare. These documents should be considered as authoritative references.

EXAMPLE
Animal health emergency response goals and strategies

For each disease listed in Australia's Emergency Animal Disease Response Agreement, a specific response strategy has been developed and documented in the form of a 'disease strategy' manual. These documents set out the nationally agreed policy (and supporting technical information) for the response to an incidence, or suspected incidence, of disease in Australia.

Response strategies include information on:
- the nature of the disease;
- principles of control and eradication; and
- policy and rationale (the agreed Australian response policy and strategies for its implementation).

Source: Animal Health Australia. 2021. AUSVETPLAN: Overview (version 5.0)https://animalhealthaustralia. com.au/ausvetplan/

Many factors need consideration when determining whether a particular strategy would be appropriate and advantageous to support the response to an animal health event. This includes, among others, the option to use a medical treatment versus a vaccine, to choose the level of intensity for the outbreak investigations, the direct and indirect consequences of the outbreak (such as animal welfare), the stakeholders' response measures and public acceptance of that, the scale of the event, the available veterinary countermeasures, and the resources available to implement selected response strategies (APHIS, 2017).

Disease response strategies should be determined in peacetime and can include the implementation of:

- movement restrictions for animals and animal by-product;
- quarantine of premises;
- stamping out; and
- emergency vaccination.

These disease response strategies are not mutually exclusive and may need to be implemented simultaneously combined with other sanitary measures. This is discussed further in Chapter 3.

2.3 ESTABLISHING RESPONSE CAPABILITIES

In addition to the prerequisites, the achievement of a state of preparedness for the management of operations during an animal health emergency requires the establishment of some specific capabilities at the country level, which include:

- authorities and enabling legal frameworks;
- partnerships and agreements;
- emergency response structure and systems;
- emergency operations centres;
- emergency operations plans and supporting procedures; and
- systems to support emergency operations.

These capabilities and how they are applied in many countries are described below.

2.3.1 Authorities and enabling legal framework

This capability consists of the ability of the national and sub-national jurisdictions to establish the legal and regulatory infrastructure to be activated during all phases of an animal health emergency. The objective is to have in place the authority and the procedures to declare an animal health emergency and execute the necessary powers to respond to the emergency in a prompt and effective manner. This is an essential element in the emergency planning process, which includes the following actions:

- establish authority and powers to act in an emergency; and
- identify and designate the emergency response authority.

2.3.1.1 Establish authority and powers to act in an emergency

The key component of animal health emergency preparedness is the availability of legal powers to carry out all necessary disease control actions. Following international standards (notably WOAH's Terrestrial Animal Health Code), and existing national frameworks (such as national disaster management frameworks and national emergency preparedness and response frameworks), the legal framework should set up the requirements and regulatory powers to determine how the veterinary services of a country and other public agencies are involved in the management of operations during an animal health emergency. This legal framework should include, but is not limited to, the following authorities and powers:

- ability to declare an animal health emergency;
- ability for officials or designated people to gain access to farms, abattoirs or other livestock enterprises to carry out emergency tasks (e.g. disease investigations, implementation and control of quarantine, culling, disinfection, sample collections, etc.);
- ability to declare infected areas and disease control zones;
- ability to quarantine farms or other livestock enterprises to prevent movement of animals and animal products both into and out of the affected premises without express permission of the regulatory authority;
- ability to ban the movement of livestock, livestock products or other potentially contaminated materials;
- ability to have access to animal movement records;
- ability to seize and obtain custody of animals;
- ability to have authority over the fate of the affected animals and decision-making power on how affected animals will be managed (e.g. select the method by which affected animals could be destroyed);
- ability to authorize the compulsory destruction and safe disposal of infected or potentially infected animals and contaminated or potentially contaminated products and materials (subject to compensation, see below);
- ability to access available finances, donations and emergency funds to support emergency actions including deployment of personnel, purchase and mobilization of equipment and materials, and proper management of compensation schemes for farmers (Not all countries have the capabilities to use the same mechanisms to finance an emergency response. Some veterinary services rely on internal or external funding to accomplish their work, while others rely on the intervention of international organizations.); and
- ability to trigger pre-emptive and precautionary interventions and implementation of disease control actions, including movement restrictions and compulsory vaccination during the alert phase.

The legal and regulatory framework should be applicable to all appropriate levels, from the national level to the sub-national and local level, according to the administrative division of the country.

For countries that operate under a federal government system, there should be harmonization and consistency under the legislations for animal health emergencies throughout the country.

All key stakeholders involved in the response to animal health emergencies within a country should be trained in the legal frameworks of their jurisdictions, in order to have a clear understanding of the legal basis for their actions.

The legal authority to declare an animal health emergency is a fundamental tool in animal health emergency management. There are several factors to consider when declaring an animal health emergency, including the risk to public health, the economic impact on the livestock sector, the impact on food security and safety, animal welfare and the impact on tourism. National and sub-national authorities need to pre-define

EXAMPLE
Legal frameworks for animal health emergencies

The Animal Health Protection Act in the United States of America gives broad authority to the Secretary of Agriculture to prevent, detect, control and eradicate animal diseases and pests, and to declare an animal disease emergency.

The Code of Federal Regulations provides guidance and authority to the United States Department of Agriculture to cooperate with states in the control and eradication of the disease. This includes:

- operations and indemnity for quarantine restrictions and orders;
- purchase, destruction and disposition of infected or exposed animals and related animal products and materials; and
- disinfection of premises, conveyances and materials, and other operations.

In addition to this, each state has the power to adopt additional statutes, rules, policies, procedures and other legal frameworks that complement the overarching federal requirements.

Source: USDA

EXAMPLE
Criteria for the declaration of animal health emergencies

- a priority disease absent from the country is confirmed in a neighbouring country;
- suspicion of the first case of a priority disease absent from the country;
- confirmation of the first case of a priority disease absent from the country;
- significant increase (thresholds) or abnormal evolution (hosts, areas) of the incidence or virulence of a priority disease present in the country; and
- high and concomitant frequency of decreased performance with no known cause in many epidemiological units (thresholds).

Source: FAO. A guide to compensation schemes for livestock disease control http://www.fao.org/ag/againfo/resources/documents/compensation_guide/introduction.html

the criteria, triggers and thresholds that will need to be met before declaring an animal health emergency and, if necessary, request external support.

Procedures for the declaration of an animal health emergency and conditions for its renewal need to be pre-established. The declaration should include the following information:

- the effective dates the declaration began;
- geographic areas covered;
- conditions giving rise to the animal health emergency; and
- the agency or agencies leading the response activities.

Declarations can also include mitigations that might be required to reduce the impact of the outbreak, such as enhanced biosecurity measures or restricted movement of affected animals and animal products.

The declaration of an animal health emergency creates an enabling and transparent situation to request additional resources that are commonly unavailable during non-emergencies. It should provide the relevant parties with the:

- authority to activate emergency response structures, plans and mutual aid agreements;
- access to additional resources including emergency funds to help respond to and recover from the emergency;
- authority to use emergency funds and deploy personnel, equipment, supplies and stockpiles;
- statutory immunities and liability protections for those involved in response activities;
- arrangements for streamlining administrative procedures such as procurement requirements; and
- emergency authorization to use medical products.

Emergency declarations can be issued at the local, sub-national and national levels.

2.3.1.2 Identify and designate the emergency response authority

The competent authority responsible for managing operations during an animal health emergency should be identified within the legal framework and should have responsibility for developing emergency plans during peacetime and the ability to mobilize needed resources during the emergency phase. This competent authority maintains the capability to respond to and manage operations during an animal health emergency within its jurisdiction. The competent authority may trigger predetermined immediate and preemptive actions before the declaration of an animal health emergency. The procedure and criteria for the delegation of authority and responsibility during an animal health emergency should be well-developed and included in the legal framework.

In most countries, the control and eradication of livestock diseases of national significance are primarily the responsibility of national governments and they are usually led by the national veterinary services. Veterinary services can be delegated by legislation or regulation as the national emergency response authority by the country's national government, with the Chief Veterinary Officer (CVO) or equivalent person or entity responsible for all technical aspects of animal health emergency preparedness and response.

In some countries, each state, territory or province may have statutory responsibility for the management of operations during an animal health emergency within its borders and may receive support from the national governments under the national legal framework. In such cases, the governmental authorities should work together to orchestrate operational procedures that achieve the response goals and objectives.

Countries may also need to establish a regulatory authority or formal mechanisms to interact with other governmental departments or ministries, which are needed to assist the national veterinary services during an animal health emergency. This includes other departments or ministries of disaster management, law enforcement, public health, wildlife, finances, environmental health, and others, including international organizations (FAO, 2021).

2.3.2 Partnerships and agreement

Following the establishment of the authorities and legal frameworks described above, the next step is to establish partnerships and agreements that will support the management of operations during an animal health emergency. Collaboration with ministries or departments, other than the national veterinary service or the ministry of agriculture, provides a platform for open and consistent communication among technical experts, which encourages knowledge and information about the emergency to be shared. To achieve an effective collaborative approach, the competent authority and stakeholders should be identified and assigned their respective role, responsibility and mandate.

These stakeholders can be identified in other governmental agencies, non-governmental organizations, the private sector and academia. Stakeholders should be invited to participate in networks and forums that are responsible for preparing for animal health emergencies. Traditional partners typically include:

- public health sector, where the animal health event presents potential human health and well-being implications. It is crucial to establish key working partnerships with the ministry of health and public health agencies for managing zoonotic diseases;

EXAMPLE
Public health collaboration

The veterinary service in Chile has a system of focal points for communication within the ministry of health to identify information on human cases (i.e. locations) during outbreaks of zoonotic diseases. It consists of subject matter experts that maintain constant communication between the ministry of health, health centres, hospitals and the national veterinary services during an emergency.

- environmental services, to provide advice on potential environmental impacts of disease response activities. in some countries the environmental services may also act as a responding department if the disease impacts wild terrestrial and aquatic animals;
- disaster management authority, which will provide additional support if the animal health emergency exceeds the capability and capacity of the veterinary services;
- private sector/industry, as they are the most knowledgeable about their sector/industry and can facilitate the implementation of response strategies;
- law enforcement authorities (police, armed forces, border security services), to support the enforcement of the response strategies, which may include control of borders to prevent the entry or exit of high-risk commodities;
- wildlife agencies or ministries of wildlife conservation. very often wildlife presents a potential reservoir for disease agents that constitute a risk for spill-over to domestic animal populations during an animal health event;

- community and cultural leaders who play an important role in communicating with communities, cultural groups and individuals;
- civil society organizations, non-governmental organizations, volunteers and research institutes, and academia; and
- international organizations that play an important role in assisting countries during an animal health emergency. Through international assistance, countries can request technical, logistical and/or operational support when required.

EXAMPLE
International organization assistance

FAO's Emergency Management Centre for Animal Health (EMC-AH) based at FAO headquarters in Rome, Italy, has been leading global animal health emergency management since 2006.

EMC-AH assists countries in managing animal health emergencies by providing guidance on emergency preparedness and response capacity.

EMC-AH also identifies needs to be addressed for improved prevention, detection and recovery after an animal health emergency.

For more information see http://www.fao.org/emergencies/how-we-work/prepare-and-respond/emc-ah/en/

When identified, the roles of key stakeholders should be documented, and formal partnerships established using agreements such as memoranda of understanding (MoU). Specific roles should be based on the types of resources and specific skills that stakeholders can provide for managing operations during an animal health emergency. These could include:

- Minimize the risk of animal health emergency occurrence by developing biosecurity plans.
- Report events likely to trigger an animal health emergency.
- Maintain adequate resource capacity (i.e. emergency response personnel, equipment, laboratory diagnostics) to respond to an animal health emergency.
- Participate in decision making during an animal health emergency.
- Share eligible response costs (i.e. responder's labour, depopulation, cleaning and disinfection costs) of an animal health emergency.

The engagement of the private sector/industry stakeholders in animal health emergency preparedness and response, especially through public-private partnerships can minimize the risk of occurrence of emergencies and enables quick and effective responses if an animal health emergency should occur. The Emergency Animal Disease Response Agreement in Australia, described below is an example of a government-industry partnership for collectively and significantly increasing the country's capacity to prepare for and respond to an animal health emergency.

EXAMPLE

Public-private partnerships at country level: Australia

Emergency animal disease response agreement

Members: 23 (Australian federal government, state and territory governments, livestock industries).

Funding: Annual subscriptions for each member calculated on the gross value of production of the industry or jurisdiction.

Strategic priorities
- Manage and strengthen Australia's emergency animal disease response arrangements.
- Enhance the emergency animal disease preparedness and response capability of Animal Health Australia and its members.
- Strengthen biosecurity, surveillance and animal welfare to enhance animal health, and support market access and trade.
- Deliver member value, organizational performance enhancement and sustainable resourcing.

When established, public-private partnerships and their documented agreements should describe agreed operating principles, guidelines and the roles and responsibilities of all parties. Further information on public-private partnerships can be found in the following:
- the WOAH Public Private Partnerships (PPP) Handbook (WOAH, 2019), which provides guidelines for development of impactful and sustainable public-private partnerships in the veterinary domain.
- the FAO Strategy for Private Sector Engagement, 2021-2025, which reflects a new forward-looking vision for strengthening strategic engagement with the private sector towards achieving the UN's Sustainable Development Goals.

A public-private partnership can allow countries to create an emergency animal health fund to be released during the emergency response or to develop an agreement between the government and private industry on how to manage the cost and responsibility for an emergency response to an animal health emergency. The public-private partnership agreements should be formal with the potential to legally bind all parties. This will allow funds to be mobilized quickly to assist the veterinary services when managing operations during an animal health emergency, while minimizing uncertainty over funding arrangements.

In addition to public-private partnerships, mutual aid agreements can be established between states or countries to ensure mutual support when the available national capacity is overwhelmed. Mutual aid can be extended to equipment, personnel, outbreak investigation, epidemiological analysis, culling and disposal of carcasses, and collaboration for laboratory support.

EXAMPLE

Public-private partnerships at country level: Namibia

A public-private partnership allowed the development of an emergency animal health fund in Namibia, which was mobilized during a Foot-and-Mouth disease (FMD) outbreak in 2015. As an emergency response, the Meat Board of Namibia (MBN) could quickly mobilise funds to assist the Directorate of Veterinary Services (DVS) to immediately set up disease control measures (procurement of control equipment and material). Through the platform of the Animal Health Consultative Forum, of which the MBN is the secretariat, the MBN also assisted the DVS through:

- awareness campaigns country-wide;
- the appointment of expert consultants in disease control and diagnosis, appointing and coordinating veterinarians to conduct post-vaccination sero-surveys;
- provision of rations to temporary staff manning road-blocks;
- coordinating, via the farmers associations, the assistance of farmers bordering the Veterinary Cordon Fence to patrol, maintain and repair the fence where necessary (continuously assisting DVS with repairing and maintenance of the fence in areas where elephant movement regularly occur).

Source: WOAH. Emergency and Resilience https://www.woah.org/en/what-we-offer/emergency-and-resilience/

Mutual aid agreements should clarify the legal authority under which the agreement is undertaken and outline procedures for requesting and providing resources.

Emergency specialists can also form agreements to provide technical expertise to the country in the development, harmonization and maintenance of the national capability and capacity to prepare for and respond to animal health emergencies.

EXAMPLE

Mutual aid agreements

The International Animal Health Emergency Reserve agreement, which includes Australia, Canada, Ireland, New Zealand, the United Kingdom of Great Britain and Northern Ireland and the United States of America, is an example of a mutual aid agreement.

The agreement permits signatory countries to share personnel in the event of an emergency animal disease outbreak, in order to supplement their domestic emergency response capabilities.

2.3.3 Emergency response structures and systems

The ultimate responsibility for the control of animal health emergencies lies with the government, and this responsibility is often devolved to the ministry responsible for animal health. In peacetime and by legislative provisions, the competent authorities should establish a clear emergency response structure and an incident command system that include a chain of command with defined roles, responsibilities and duties for the management of operations during an animal health emergency. The response structure should allow the CVO or designated official to effectively and efficiently manage any priority animal health event from the national central level through to the local field level.

The development of an emergency response structure for the management of operations during an animal health emergency should take into consideration two important emergency response roles:

1) an on-scene response management role, which oversees the field activities; and
2) an off-scene emergency coordination role, which oversees the general coordination of the response operations.

The emergency response structure should be scalable and adaptable to ensure efficient and effective responses to animal health emergencies of any size and complexity. It should also be sufficiently flexible to quickly expand as additional external resources are added to match the increasing demands of the animal health emergency, and to readily contract as the emergency phase transitions into reconstruction.

EXAMPLE
Emergency response structure

The United States Department of Agriculture has established an emergency response structure that is used in animal disease incidents. Within this structure, the Administrator of the United States Department of Agriculture's Animal and Plant Health Inspection Service (APHIS) is the federal executive responsible for implementing APHIS policy during an outbreak.

Many incident management functions are delegated to the Veterinary Service Deputy Administrator, who is the CVO of the United States of America.

This organizational structure, which is illustrated in Figure 2, handles requests for resources, coordination and policy from the teams operating in the field (AHA, 2021).

The emergency response structure should incorporate an incident management system that can be applied at all levels. Contemporary incident management systems are based on the Incident Command System (ICS), which was introduced in North America in 1970 to manage devastating wildfires. Many countries have adopted and adapted this to their specific needs and its use has been extended to materials spills, earthquakes, human health, animal health emergencies and planned events (such as the Olympic Games). The primary management functions of an incident management system include:

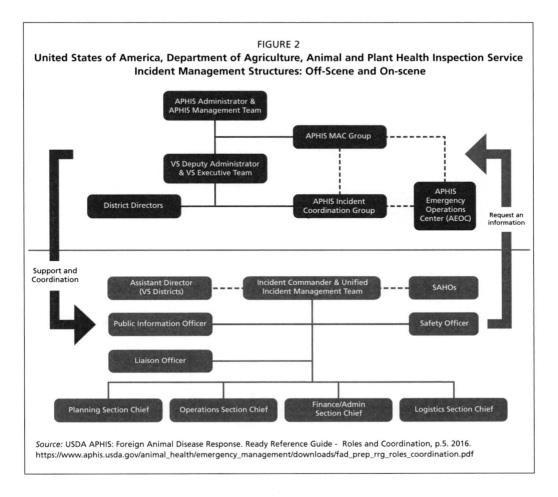

FIGURE 2
**United States of America, Department of Agriculture, Animal and Plant Health Inspection Service
Incident Management Structures: Off-Scene and On-scene**

Source: USDA APHIS: Foreign Animal Disease Response. Ready Reference Guide - Roles and Coordination, p.5. 2016.
https://www.aphis.usda.gov/animal_health/emergency_management/downloads/fad_prep_rrg_roles_coordination.pdf

- incident Command;
- operations;
- planning;
- logistics; and
- finance/Administration.

These functions and their relationship are discussed further in Chapter 3.

It is important to note that responders can perform multiple roles within an incident command system and that all functions do not have to be activated at the same time. For a small incident, for example, the same person may be responsible for both logistics and operations, while another person may oversee the finance/administration and planning functions for the incident.

The advantage of a modular structure is that it can be expanded to include more functions and levels to meet the needs of the response. The levels and numbers of functions activated will depend on the scope, complexity, disease or agent, geographical area and size of the response required. The structure may also be contracted or, as the incident ends, collapsed, and moved towards demobilization and recovery. This is discussed in further detail in Chapter 3.

2.3.4 Emergency operations centres

This section presents the core features for establishing in peacetime an emergency operations centre and its role in the management of operations during an animal health emergency.

The World Health Organization's (WHO) Framework for a Public Health Emergency Operations Centre (PHEOC) describes the core components of an emergency operations centre as the legal authority, the physical infrastructure, the plans and procedures, the information and communication technology (ICT) infrastructure, the information systems and data standards, the human resources, the training and exercises, the monitoring and evaluation and the financial resources (WHO, 2015).

The efficient management of operations during an animal health emergency requires effective coordination of all parties involved. Support and off-scene coordination should be provided from a pre-designated and prearranged emergency operations centre, which may be as small as a single vehicle, a desk, a conference room, or as large as an entire building.

2.3.4.1 The role of the emergency operations centre

The emergency operations centre is critical to managing operations during an animal health emergency. It manages all off-site activities and provides a central location for responders to gather and coordinate resources and requests for on-ground activities.

During an animal health emergency, the emergency operations centre may:

- provide direction and support for on-ground activities;
- collect, collate, evaluate and disseminate information;
- coordinate off-site agencies and operations;
- prioritize resources;
- manage resources (needs, requests, allocation and tracking); and
- manage public information and warnings.

Depending on administrative divisions and available resources, countries may decide to establish emergency operations centres at national, sub-national and local levels. While the focus of each centre can be different, the structures, systems and processes established in each will have many similarities.

2.3.4.2 Establishing an emergency operations centre in peacetime

Establishing an emergency operations centre begins with defining the functional requirements and should address the planning steps outlined in Figure 3.

Further guidance on establishing an emergency operations centre is provided in Annex 2.

2.3.4.3 Emergency operations centre policies and procedures

The emergency operations centre should have in place written policies and standard operating procedures (SOPs) to ensure effective running of the emergency operations centre and the management of operations for an animal health emergency. These SOPs need to be developed in peacetime, to ensure they are available in the emergency phase and include:

FIGURE 3
Planning steps for the establishment of an emergency operations centre

1. Perform a situation analysis

Identification of the incident based on risk assessments

For animal health emergencies, the types of threats will most likely include the occurrence of high consequence animal diseases and pests

2. Determine the staffing requirements

Types and number of essential personnel required

The number of necessary staff should reflect the level of activation and work demands of the animal health emergency.

3. Perform an EOC operational requirements survey

Background information

This information will provide guidance in terms of the location, relocation, and site selection of the emergency operations centre.

4. Determine space requirements

Circulation and construction layout requirements, expansion requirements

The emergency operations centre may have dual uses during non-emergency conditions, provided that it can be rapidly converted to accommodate the functions necessary to respond to the animal health emergency.

5. Perform a security risk analysis

Controlled access area, survivability, scalable security

The emergency operations centre should be a controlled access area, with safety of personnel and information integrated into planning.

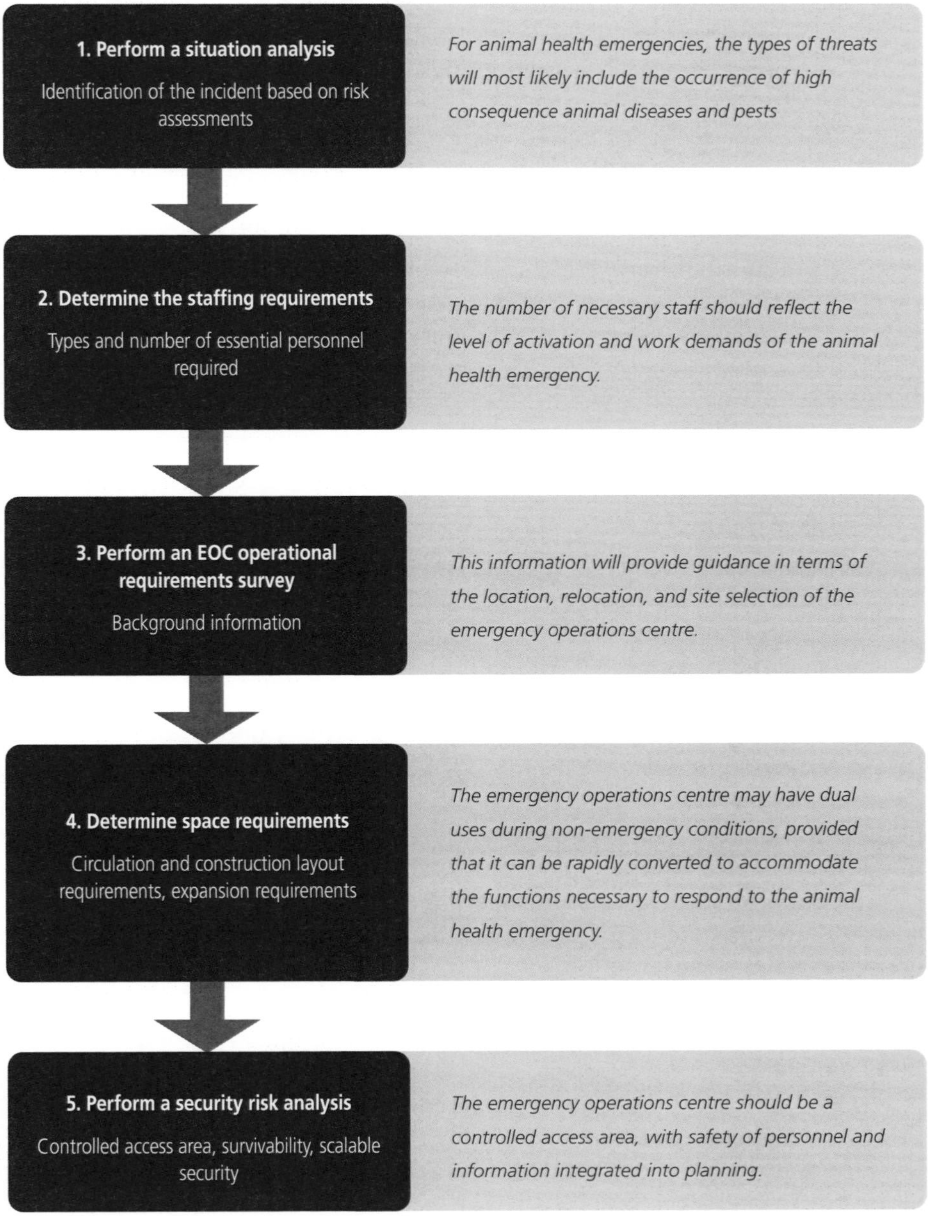

Source: FAO, 2022.

- triggers for activation of the emergency operations centre (i.e. alert notifications, trends in disease surveillance data, emergency declaration, etc.);
- procuring and processing of contracts during an animal health emergency;
- preparing and processing of reports;
- establishing and maintaining life support systems, within the emergency operations centre (i.e. accommodation, food services, water, sanitary facilities, medical supplies, heating, ventilation, air conditioning);
- operating equipment; and
- management of records and documentation.

The decision-making process for both activation and deactivation of the emergency operations centre should also be documented in an SOP and should clearly indicate:

- **who** has the authority to make the decision;
- **what** are the circumstances for activation or deactivation;
- **when** should the activation or deactivation occur; and
- **how** is the level of activation or deactivation determined.

The level of activation or deactivation of an emergency operations centre should be based on established triggers.

2.3.4.4 Equipping the emergency operations centre

When preparing for an animal health emergency, actions should include the development of anticipated lists with standard specifications, acquisition of the standard equipment and expendable supplies (such as personal protective equipment [PPE]) that the organization may need to use to effectively manage operations during an animal health emergency.

A checklist of equipment and supplies for an emergency operations centre is included in Annex 1.

2.3.5 Emergency operations plans and supporting procedures

Management of operations during an animal health emergency requires the development of a library of plans and related SOPs. These plans and SOPs should be adapted to the country's available capacities and the local context. Their functionality, effectiveness and readiness should be tested through simulation exercises and reviewed regularly to make sure that they are fit for purpose. An animal health emergency library can include:

- emergency response plans;
- compensation plan;
- emergency recovery plan;
- continuity of business/operations plan; and
- SOPs.

2.3.5.1 Emergency response plans

Emergency response plans (also known as contingency plans, or emergency intervention plans) refers to a document or series of documents used during emergency phase of an animal health emergency. These plans can include the relevant measures, concept of operations, procedures, information and tactics that should be implemented to manage operations during an animal health emergency.

Examples of emergency response plans (for both aquatic and terrestrial animals) are available from the WOAH Emergency and Resilience website (WOAH).

2.3.5.2 Compensation plan

Animal health emergencies can have large economic consequences for farmers in terms of direct and consequential losses. The purpose of the compensation plan is to describe the agreed policy and procedures for the compensation of animal owners in case of compulsory culling and destruction during an animal health emergency.

The compensation plan should be informed by a country's compensation policy. Details of compensation, such as the level of compensation paid, how it is paid and when it is paid, should be considered during peacetime and documented in the compensation plan.

2.3.5.3 Recovery plan

The recovery plan addresses short- and long-term recovery priorities. The plan provides guidance for restoration of identified critical functions, services/programmes, vital resources, facilities and infrastructure to the affected area.

2.3.5.4 Continuity of business plan

The continuity of business plan aims at mitigating the consequences of an animal health emergency by allowing veterinary services and affected businesses to continue to perform normal operations or critical activities during an emergency response.

2.3.5.5 Standard operating procedures for management of operations

SOPs provide details for conducting critical activities that are essential to effectively manage operations during an animal health emergency. They provide operational details that are not discussed in depth in the disease strategy documents or emergency response plans.

They should be adopted (and possibly developed) at the local/operational level. SOPs should be made available at the location where the work is done and be included in the training package of the stakeholders and responders involved in the management of operations during an animal health emergency.

Different formats may be used for SOPs depending on the intended audience and purpose. However, each SOP should include the following items:

- title;
- purpose or rationale statement;
- authority signature(s);
- application/scope;
- resources/equipment;
- warnings;
- description of activities and who is authorized to conduct each activity;
- references; and
- appendices (where applicable).

Examples of SOPs for emergency response to Highly Pathogenic Avian Influenza (HPAI) can be found on the Good Emergency Management Practice, Standard Operating procedures for HPAI Response on the FAO website (FAO, 2011).

2.3.5.6 Maintaining plans and standard operating procedures

Emergency response plans and SOPs are not static documents and procedures should be established for their review and maintenance to ensure they remain up to date. The revision should take into consideration any change in appointed officials, response resources (e.g. policies, personnel, and organizational structures), management processes, and facilities or equipment. The change in hazard profiles, after-action reports and improvement plans from exercises or actual events, the enactment of new or amended laws or executive orders should also be included in the revision process. Countries should establish a quality assurance process to control changes to emergency response plans and supporting procedures.

2.3.6 Systems to support emergency operations

The optimal management of operations during an animal health emergency requires the establishment of systems and processes for managing critical activities. This section provides guidance on these systems, which includes:

- financial management systems;
- resource management systems; and
- information management systems.

2.3.6.1 Financial arrangements and systems

Funding for the management of operations during an animal health emergency is one of the critical competencies of the PVS tool, which is used to evaluate the performance of veterinary services. It includes the funds for emergency operations and compensation processes in case of destruction of animals or property. The funding arrangements established in peacetime will serve to ensure the availability and accessibility of the funds during the emergency phase. The emergency funding framework should also include procedures for the acceptance and use of external donated funds with signed pre-agreements. The designated authority and the appropriate delegations for the release of these funds should also be identified. The funds can be activated once an animal health emergency is declared and a mechanism for the rapid release of such funds is critical to prevent delays in the management of operations. Emergency funds are not necessarily under the control of the national veterinary services, and this mechanism will include the necessary procedures to release the emergency funds from the appropriate ministry or agency responsible for those funds.

Financial management systems should accommodate arrangements for managing the costs for ongoing processes including disease surveillance, risk analyses, installation of information management systems, and costs that are likely to occur during an animal health emergency, which can include:

- purchase of emergency supplies and equipment;
- transport for emergency operations personnel (and equipment);
- accommodation and meals for emergency operations personnel;
- contractors/service providers;
- establishment and maintenance of telecommunications and/or other modes of communication; and
- payment of compensation to affected farmers.

2.3.6.2 Resource management systems

Resource management systems and processes should be established and maintained to support the management of operations during an animal health emergency. These systems and processes should be flexible, scalable and adaptable, in order to support any type of animal health emergency. The GEMP manual, section titled *Prepare for an emergency in peacetime – equipping*, provides a list of activities that will assist in ensuring that resources are ready to be mobilized when an animal health emergency occurs.

EXAMPLE
Financial arrangements and systems

Most financial arrangements are country specific. For example:
- In Uganda, the spending authority is the CVO.
- In the United States of America, the regular operating budget under USDA maintains limited funds to support an emergency response. However, the Secretary of Agriculture may tap into additional funds at the Secretary's discretion and/or issue an extraordinary emergency declaration.
- In Colombia, they can redirect the monetary funds from what is called a *parafiscalidad* to another, to assist in an emergency response. The funds for the *parafiscalidades* come from mandatory monetary contributions from the producers within an agricultural sector. For example, funds from the coffee sector *parafiscalidad* can be moved into the swine health sector parafiscalidad to help with a swine disease emergency.
Colombia has 14 agricultural *parafiscalidades*, whereas Chile, for example, does not have this type of resource. Colombia modelled their *parafiscalidad* on a German system.
In Colombia, the Ministry of Agriculture and Rural Development is responsible for overseeing these funding systems and provides the policies and guidance on how to properly use these resources in different circumstances, including emergency response, as outlined on the ministry's website.

Standardized resource management systems and processes should be developed and implemented. They will facilitate the sourcing, dispatch, deployment and recovery of resources before, during and after the incident.

The management of resources during an animal health emergency is a finite process and can include the following steps as shown in Figure 4.

For the maintenance of acquired resources, a procedure should also be in place to continuously monitor the availability and condition of physical resources.

2.3.6.3 Information management system

The efficient management of operations during an animal health emergency requires information sharing between responders to facilitate planning for and implementation of

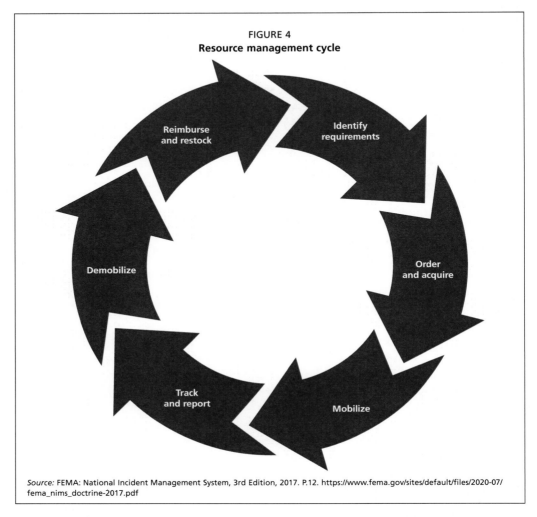

FIGURE 4
Resource management cycle

Source: FEMA: National Incident Management System, 3rd Edition, 2017. P.12. https://www.fema.gov/sites/default/files/2020-07/fema_nims_doctrine-2017.pdf

response actions. This calls for the establishment of an information management during peacetime and will require responders to be trained in its implementation.

Veterinary Services are responsible for detecting and responding to animal health events that occur within their boundaries and must cooperatively manage the data and information related to such animal health events, with many actors and partners. This requires a system that is ideally available nationwide and is flexible to support numerous data sets collected through various methods. To support this, the information management system should provide a secure accessible system for data collection, collation, analysis and dissemination. Such an information management system could include case management, investigation, tasking, resource management and geographic information system (GIS) software, which sits within a universal information platform, which is accessible to approved users.

The following factors should be considered when establishing an information management system:

> **EXAMPLE**
> **Information management systems**
>
> The veterinary emergency response system in the United States of America is support-ed by the Emergency Response System 2.0 (EMRS 2.0), a web-based application used for reporting and storing disease surveillance data, data from state-specific outbreaks and national (all hazards) emergency responses, and data from routine foreign animal disease investigations.

- the incorporation of stakeholders into the information flow with their respective roles and agency mandates;
- the rules for the collection, collation, interpretation, dissemination, access and retrieval, storage and retention, and security of the information;
- the confidentiality guidelines for the information;
- the training of relevant stakeholders and users; and
- the means, channels, technologies, equipment and tools to use.

2.3.6.4 Operational communication systems
Briefings, debriefings and reporting are essential components of the operational communication process and critical to information flow during an animal health emergency.

Briefings ensure all personnel involved, including contractors, understand the objec-tives, strategies, safety issues, roles and responsibilities, and reporting relationships. They occur at activation, change of shift, and when the team enters and disperses in a survey area. Briefings are expected to be short and to the point. A briefing is not a discussion, an opportunity for debate, or a question and answer session. Briefings are given by those in senior positions – the incident manager is ultimately responsible for ensuring that brief-ings occur at all levels of the incident management structure, and that planning, public information, operations and logistics conduct appropriate briefings within their sections.

Briefings should follow a predefined and standardized format like the one used in the Australian Incident Management Systems (SMEACS-Q). This acronym is a useful short-hand way to remember how to brief, and how briefings are conducted. This is important for the informant and the audience and provides a routine and expected structure to the information flow. A checklist for the SMEACS-Q briefing approach is provided in Annex 1.

2.4 MAINTAINING RESPONSE CAPABILITIES
The effort that is put into establishing response capabilities by veterinary services must be retained and where possible, improved over time. Common approaches to maintaining and improving response capability include:
- training of personnel;
- testing through simulation exercising; and
- monitoring and evaluation.

2.4.1 Training of personnel

Stakeholders that have a responsibility to implement or manage operations during an animal health emergency should receive training on their responsibilities, from executives to field personnel. Perceived and actual training needs among personnel with and without experience should be identified, in order to promote a continuous capacity building process. The core competencies vary among the emergency responders, and the competencies required for tactical operations (on-site) and emergency response management (leadership and coordination) personnel are not the same.

The on-ground response personnel require technical skills to manage on-ground activities. These may include sanitary measures, such as depopulation, disposal, decontamination, quarantine and movement control, the use of equipment and tools, risk communication and behaviours.

The emergency response management personnel need to have technical and non-technical skill sets for addressing managerial tasks in support of operations.

Training programmes can range from multi-year sessions that are required for incident management teams and individual response team members, to a broader set of core training requirements designed to track and verify the competencies of incident management and response team members in order to ensure they can fulfil their specific roles in an animal health emergency.

The training programme should also include just-in-time training materials for refreshing or training incident management teams and responders, including volunteers or day labourers, so they can adequately and safely fulfil their assigned position during an animal health emergency.

2.4.2 Testing through simulation exercises

An exercise is a focused practice activity that can place teams or players in a simulated situation. During an exercise, the teams or players may need to function in the capacity that would be expected of them in a real event.

Exercises can be conducted to support an overall assessment of the response capabilities, mentioned above. There are two main benefits to conducting exercises:

1) Individual and team practice: Exercising enables people to practice and gain experience in their roles.
2) System improvement: Exercising can assist in improving an organization's system for managing operations during an animal health emergency.

Benefits come from participating in, planning, controlling and evaluating the exercise, and acting upon lessons identified during the exercise. Exercises can be conducted to:

- test the functionality of emergency response plans and SOPs;
- demonstrate the proficiency of incident management teams;
- demonstrate or practice the proper use of equipment and supplies; and
- practice and confirm the capacity and practical skills of the competent authorities and stakeholders.

2.4.3 Monitoring and evaluation in peacetime

Monitoring and evaluation of the preparedness activities, described in this section, should be undertaken during peacetime by the competent authority to ensure that preparedness activities are current and reflect contemporary practices. For this reason a cycle of monitoring and evaluation should be implemented for:

- agreed arrangements, such as legislation, agreements, policies, plans and procedures;
- resources to be used in an animal health emergency, such as facilities, stockpiles and response equipment; and
- response personnel, to ensure their response skills and knowledge are adequate for responding to an animal health emergency.

While exercises are an acceptable way to evaluate how prepared a country is to manage operations during an animal health emergency, exercise should also be supported by a formal program of monitoring and evaluation that ensures that lessons are identified and in turn contribute to continual improvement in a country's level of preparedness for managing operations during an animal health emergency.

Chapter 3
Managing operations during an animal health emergency

3.1 INTRODUCTION

This section of the manual covers the actions to be taken during the alert and response phases of an animal health event, which could result in an animal health emergency. The effectiveness of the response to an animal health emergency is dependent on an early triggered alert, the quality of the information collected during the alert phase and the subsequent actions taken during the emergency phase.

3.2 ALERT PHASE

The alert phase refers to the period of time when the level of risk due to an animal health event requires close observation of all activities, rapid transmission, sharing and assessment of relevant information and quick precautionary action to address an impending emergency.

The alert phase is the period when a threat is coming closer or has been identified, such as the suspicion of a case of a priority disease or confirmed outbreaks in the vicinity of, or that threaten the country. During the alert phase, early warning systems may be used.

3.2.1 Investigating suspicion of an animal health event

The alert phase begins when the national or local veterinary authority receives information suggesting the occurrence of an animal health event which has the potential to become an animal health emergency. Information can be shared by private or government veterinarians, laboratory staff, farmers or producers, industry representatives and community workers, and may come from a number of sources, including:

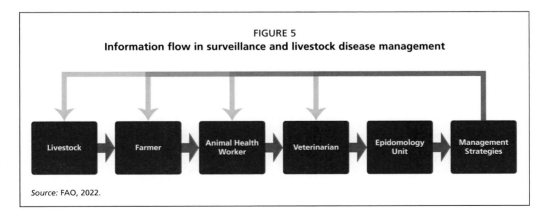

FIGURE 5
Information flow in surveillance and livestock disease management

Livestock → Farmer → Animal Health Worker → Veterinarian → Epidomology Unit → Management Strategies

Source: FAO, 2022.

- routine surveillance monitoring, such as inspection at slaughterhouses; and
- epidemiological investigations.

After receiving the information, the national or local veterinary authority is responsible for ensuring that an investigation is conducted using a standardized procedure and forms. This involves a situational assessment based on inspection of affected animals, history taking and specimen collection to determine the nature and extent of an event. This assessment should be re-evaluated on a continuous basis. Figure 5 illustrates the typical flow of information in surveillance and livestock disease management.

Following the initial investigations, it is the responsibility of the national veterinary authority to decide whether the report constitutes a suspicion. This decision must be based on a collection of clinical and epidemiological information and technical expert opinion.

When an impact on public health is suspected, the investigation should include the public health authority.

3.2.2 Confirmation of the suspicion

According to the procedures and the legal framework defined in peacetime and taking into account the results of the field investigations and the history of the event information, the confirmation of a suspicion may be based on confirmed diagnostic results of a reference laboratory (or approved laboratory), or in some circumstances, clinical signs of strong suspicion.

EXAMPLE

Confirmation of suspicion

In Canada, confirmation of the index case of an animal health disease can only be done by the National Centre for Foreign Animal Disease (NCFAD) of the Canadian Food Inspection Agency (CFIA).

The case definition, as well as the identification of the reference or approved laboratory play a key role in the confirmation of the index case. The responsibilities of laboratories in determining the case definition should be defined during peacetime. The laboratory team may be required to join the investigation team in the field and support the collection of the samples. The laboratory team should immediately notify the CVO if a sample result is suspected or confirmed positive.

Using all the information and tools available, including the case definition in the response plans, it is the role of the veterinary authority to determine if the situation meets the established criteria for confirmation of an animal health emergency. If it is a zoonotic disease, the information should be shared with the public health authorities to involve them in the emergency response.

The following box provides examples of observations that may trigger joint investigations with public health authorities.

EXAMPLE
**Observations that may trigger joint investigations
with public health authorities**

- single cases of zoonotic diseases critical to a particular sector based on international regulations;
- an unusual signal or unexpected trend in surveillance data or analyses of health indicators;
- reported through sector-specific or a coordinated surveillance system or other system for early warning;
- rapid or complex political, social or economic change, man-made or natural disaster;
- declaration by WHO of a public health emergency of international concern;
- reports to the WOAH of a confirmed zoonotic disease outbreak in animals;
- notification from the International Food Safety Authorities Network secretariat regarding a zoonotic food safety issue; and
- new perceptions, for example from social media, of government concern or international or non-governmental organization statements.

Source: FAO. A guide to compensation schemes for livestock disease control http://www.fao.org/ag/againfo/resources/documents/compensation_guide/introduction.html

3.2.3 Response actions to implement during the alert phase

Initial response measures may be needed prior to the official confirmation of the animal disease. The objectives of response actions during the alert phase are to avoid the spread of a pathogen if it were to be confirmed, minimize the reaction time and enhance effectiveness of actions during the emergency phase. In some countries performing actions before a declaration of an emergency can be difficult. In peacetime, this provision for the initiation of response actions should be enabled by the emergency management legal framework.

Based on the results of the investigations, the veterinary authority may decide to initiate disease control activities either when the disease is under suspicion, if there is an absence of information on the extent of the outbreak (from field surveillance), or if there are unknown features of the disease agent. Figure 6 identifies activities that may be implemented in the alert phase, prior to the formal declaration of an animal health emergency.

During the alert phase, legal instruments, such as a declaration of emergency, declaration of a movement standstill, classification of the different areas and movement or disease control measures, etc., may need to be implemented to ensure they are suitable for the specific animal health emergency.

FIGURE 6
Activities to be implemented in the alert phase

Movement controls

- Restrictions of animal movements in a defined area
- Temporary control zone with surveillance of farms in the zone
- Quarantine restrictions

Coordination

- Activation and standby of responders and coordination mechanisms
- Stakeholders alerted and briefed
- Increase collaboration efforts between national and sub-national jurisdictions to achieve response goals.

Communication

- Drafting of materials
- Enhancing public awareness
- Dissemination of early warnings
- Increasing communication efforts between national and sub-national jurisdictions to achieve response goals.

Pre-emptive operational measures

- Field investigation
- Pre-emptive culling
- Preparatory activities for the culling (mapping, census, etc.)
- Enhanced surveillance
- Assessment and securisation or preposition of resources

Source: FAO, 2022.

3.2.4 Notification of an animal health emergency

If applicable and following the protocol developed in peacetime for the rapid and transparent reporting of suspected cases, the national veterinary authority must notify the international and regional organizations and neighbouring countries.

Cross-border meetings may occur based on the risk of disease spread to neighbouring countries. For the purposes of the WOAH Terrestrial Animal Health Code, members of the WOAH should immediately notify outbreaks of transboundary animal diseases and other high impact animal health events to international organizations, such as WOAH's WAHIS and FAO's Emergency Prevention System for Transboundary Animal and Plant Pests and Diseases (EMPRES).

At the regional level, there may also be obligations to notify regional organizations such as the African Union Inter-African Bureau for Animal Resources, the European Union, and the Association of Southeast Asian Nations.

Early international reporting will enable early access to international assistance and coordination of response, if required.

3.3 EMERGENCY PHASE

The emergency phase is the period calling for immediate and urgent actions to prevent the spread of the pathogen, mitigate direct and indirect losses caused by it and, if possible, to eradicate it. The emergency phase begins when the animal health emergency is confirmed by laboratory diagnosis, or highly suspected and presumed to exist based on clinical signs and supporting evidence from the field investigation. If the situation meets the definition of an animal health emergency, the authority in charge will declare an animal health emergency using the procedures defined and documented in peacetime.

3.3.1 Declaration of an animal health emergency

The declaration of an animal health emergency will trigger the initiation and implementation of the emergency response plan aiming at locating, isolating and controlling or eradicating the animal health emergency. According to pre-established procedures, the declaration should address the:

- effective dates of when the declaration begins;
- geographic areas covered;
- movement controls within and/or between defined zones;
- conditions giving rise to the emergency; and
- agency or agencies leading the response activities.

Having the animal health emergency officially declared before taking response actions facilitates and expedites the release and mobilization of resources in a timely fashion. This mobilization might include resources from other government agencies who are response partners.

3.3.2 Activation of the emergency response plan

The activation of the emergency response plan, pursuant to the procedures defined in peacetime, will initiate the implementation of emergency response operations and several key activities including the establishment of the coordination and command structures, a situational assessment and the implementation of control measures.

Implementation of the emergency response plan is the trigger to undertake a range of emergency response actions, which include:

- undertaking a rapid risk and situational assessment;
- activating coordination structures;
- activating the emergency operations centre;
- developing and implementing an incident action plan;
- mobilizing and deploying resources; and
- implementing emergency critical activities.

3.3.2.1 Undertake a rapid risk and situational assessment

The purpose of the initial rapid risk and situational assessment is to collect and analyse data to confirm the emergency, develop a case definition, describe its impact and possible evolution.

The adequacy of existing response capacity and immediate additional needs should also be assessed at this time, to determine the scale of operations that may need to be activated for an effective response.

The initial rapid risk and situational assessment allows officials to determine the extent and impact of the emergency and possible evolution as well as the existing response capacity and immediate additional needs. This results in the determination of the level of coordination and structures required to manage operations during an animal health emergency.

3.3.2.2 Activate coordination structure

The authority responsible for animal health emergencies should implement a coordination structure that is commensurate with the impact and possible evolution of the animal health emergency within the country. Where an animal health emergency can be managed at a local level, this may be the only element of the coordination structure that needs to be activated. However should the animal health emergency be wide-spread with significant impacts, it may be necessary to activate local, sub-national and national coordination structures.

For example, Australia's Biosecurity Incident Management System describes five levels of response for biosecurity incidents, which include animal health emergencies.

The use of these levels when describing an event or incident provides guidance on the level of coordination and resources (financial, physical and human) required to manage the response. The levels are used both within an organization and externally to communicate resource requirements to external partners. These levels are described in the Table 1.

When practical, the coordination structure should be tiered. This will allow for operations to be managed at an appropriate level. As indicated in the levels of response, a level 1 animal health emergency can be managed at a local level, utilising local resources. The local and initial response is the foundation upon which all responses are built and can be expanded to suit the needs of the animal health emergency.

TABLE 1
Example – Levels of response for an animal health emergency – Australia

	Level 1	Level2	Level 3	Level 4	Level 5
Response	Localised	Local or region (within a state)	State-wide	One or more states with insufficient resources	One or more states with insufficient resources at national level
Resources	Local with little or no external support	Local, with some support from state level	State, with possible support from other agencies or states	State, with support from national level	State, with support from national and international levels
Facilities activated for managing the response	Minimal	Dedicated local level emergency operations centre / State level coordination and support	Dedicated, one or more, local level emergency operations centres / State level emergency operations centre	State level emergency operations centre / National level emergency operations centre	State level emergency operations centre / National level emergency operations centre

EXAMPLE

Application of a tiered approach to carcass disposal

In Figure 7, the principle of a tiered response is applied to carcass removal. In this example, local government officials are the first actors on-scene and may request state assistance if local officials do not have the authority or sufficient capabilities to safely remove and dispose of animal carcasses. In a similar fashion, the state government officials may request federal assistance if state capabilities are exhausted and not sufficient. According to the country's administrative structure, coordination structures can be centralized with a relatively strict hierarchical structure or decentralized to a district or municipality, which can have relative autonomy and operate independently while communicating with the central government.

FIGURE 7
Example of tiered response in carcass removal scenario

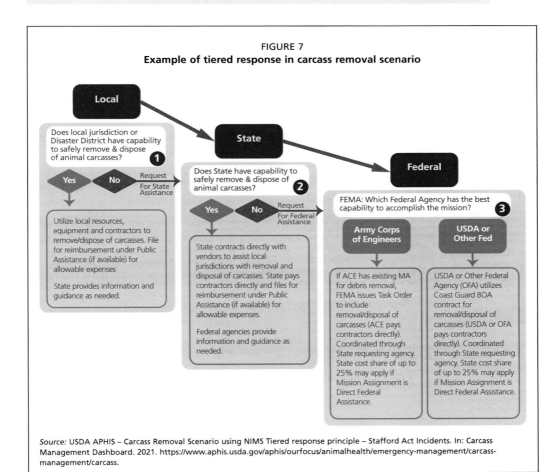

Source: USDA APHIS – Carcass Removal Scenario using NIMS Tiered response principle – Stafford Act Incidents. In: Carcass Management Dashboard. 2021. https://www.aphis.usda.gov/aphis/ourfocus/animalhealth/emergency-management/carcass-management/carcass.

3.3.2.3 Activate the emergency operations centre

The animal health emergency operations centre is the central facility with overall responsibility for the management of operations during an animal health emergency. As mentioned in Chapter 2, the emergency operations centre manages all off-site activities and provides a central location for responders to gather and coordinate resources and requests for on-ground activities.

The activation and management of an emergency operations centre for an animal health emergency should ideally be under the control of the CVO or equivalent who has legislative responsibility for all aspects of the response to an animal health emergency. Ideally the emergency operations centre should be separate from the CVO's office. This allows the CVO's office to continue its day-to-day business, as well as providing high level policy direction to the emergency operations centre.

When activated, the emergency operations centre will:

- determine and implement operational objectives and priorities;
- manage all operational activities;
- maintain situational awareness and disseminate information;
- acquire, coordinate and deliver resources required to conduct operational activities; and
- liaise with relevant stakeholders.

The responsibilities of the emergency operations centre are managed through the incident command system, which consists of five sections. These are outlined in Table 2.

A typical organisational structure for the incident command systems and sections within an emergency operations centre is shown in the following example.

TABLE 2
Sections within an emergency operations centre

Incident command	This section has overall responsibility for managing the emergency operations centre and operational activities.
	Sets objectives and priorities for the successful resolution of the animal health emergency.
Operations	This section manages the operational activities required to achieve the objectives set by the incident command.
Planning	This section undertakes response planning activities.
	Collects, collates, analysis and disseminates information about the animal health emergency.
	Plans and monitors resource requirements for operational activities.
Logistics	This section provides logistical support, obtains resources and other services required to meet operational needs. This includes purchasing, human and physical resources, transport and communications equipment.
	This section sets up the emergency operations centre and ensures it meets the needs of the animal health emergency.
Finance and Administration	This section manages financial and administrative arrangements for the animal health emergency. This may include securing funding for the emergency phase and administering contracts for the provision of services.

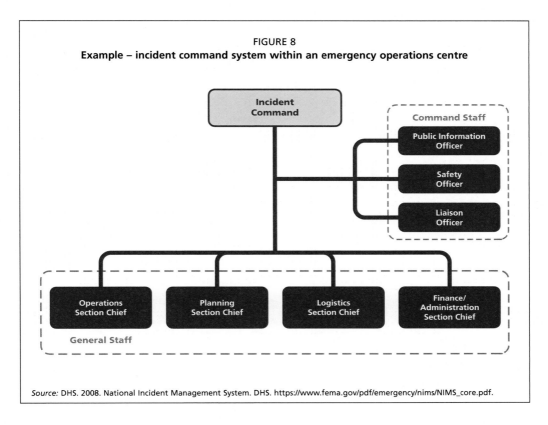

FIGURE 8
Example – incident command system within an emergency operations centre

Source: DHS. 2008. National Incident Management System. DHS. https://www.fema.gov/pdf/emergency/nims/NIMS_core.pdf.

Further information on activating an emergency operations centre and responsibilities of each section is described in Annex 2.

3.3.2.4 Develop and implement an incident action plan

One of the products of the planning section, within an emergency operations centre, is the incident action plan (IAP), which facilitates the successful management of operations during an animal health emergency.

The IAP is a living document that is updated and re-issued as the animal health emergency evolves. The format of an IAP will generally include a high-level description of the current situation and what is being done about it and how this will be achieved. The IAP should include the following information:

- the current situation (what has happened);
- Incident objectives and priorities (what do we want to achieve?);
- response strategies (priorities and the general approach to accomplish the objectives);
- response tactics (methods developed by operations to achieve the objectives);
- the resources needed for each operational period to reach the objectives;
- the operations organization necessary for the selected strategy and tactics;

- overall support including logistical, planning, and finance/administration functions; and
- health and safety plan (to prevent responder injury or illness and to provide psychological support).

A template for an IAP is included in Annex 1.

Incident action planning is a cyclical process, where the planning steps are repeated every operational period. The operational period is the length of time set to achieve a given set of objectives and is determined by the dynamics of the animal health emergency. In the early stages of an animal health emergency, it may be necessary to update and re-issue the IAP every 24 hours, however as the animal health emergency evolves and activities become more routine, the operational period may extend to several days.

The IAP is developed and approved using a planning cycle, known as the planning "P" (see Figure 9). The planning P depicts the stages in the incident action planning process that occur within each operational period.

The incident action planning process is built on the following phases:

1. Understand the situation.
2. Establish incident objectives.
3. Develop the incident action plan.
4. Prepare and disseminate the incident action plan.
5. Execute, evaluate and revise the incident action plan.

The IAP also provides an effective format and the information required for delivering operational briefings.

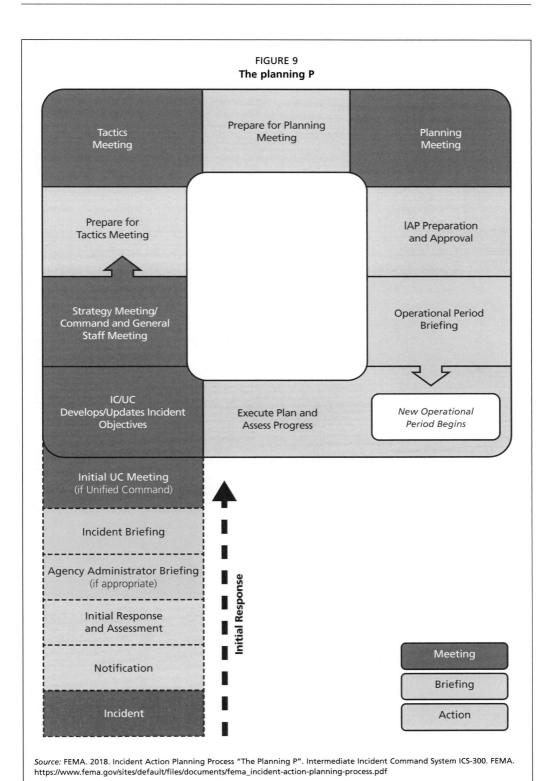

FIGURE 9
The planning P

Source: FEMA. 2018. Incident Action Planning Process "The Planning P". Intermediate Incident Command System ICS-300. FEMA. https://www.fema.gov/sites/default/files/documents/fema_incident-action-planning-process.pdf

3.3.2.5 Mobilize and deploy resources

The resources required to manage operations during an animal health emergency, such as personnel and equipment, must be mobilized, released and deployed in a coordinated manner. The logistics section in the emergency operations centre is responsible for the mobilization and deployment of resources and should implement systems and processes to order, acquire, mobilize, track, report, recover, demobilize, reimburse, restock and run inventories of the resources.

Resource mobilization and deployment can be enhanced through the use of anticipated lists of first responder equipment in conjunction with the establishment of strategic pre-positioning sites and coordination with identified contractors.

If local resources are likely to be exhausted, planning and logistics managers should tap into pre-established lists of contractors, short-term hires or volunteers. Where national level resources are likely to be exhausted, it may be necessary to activate bilateral MoUs or activate international agreements to supplement the national resources. This may include accessing FAO's EMC-AH or activating the International Animal Health Emergency Reserve Agreement or the EU Veterinary Emergency Team (European Union, 2007).

Where external resources are provided, each responder unit should include at least one person with prior training and/or experience for the assigned tasks. Just-in-time or refresher training (CFSPH) may need to be provided to responders prior to deployment. If necessary, this can be incorporated into the check-in process of the personnel.

Activation of pre-established emergency funding arrangements helps to ensure that a country will have the resources required to mount and sustain a response.

Funding arrangements may include identifying source of funding which can be public or private funds (e.g. Namibia's Emergency Animal Health Fund for Foot-and-Mouth Disease [WOAH, 2019]) or cost shared (e.g. Australia's Emergency Animal Disease Response Agreement).

3.3.3 Implement disease control measures

The implementation of disease control measures during an animal health emergency is based on four key complementary operation elements of rapid response, rapid identification, containment, control and eradication. The incident commander is responsible for ensuring these operations are undertaken to control and eradicate the disease.

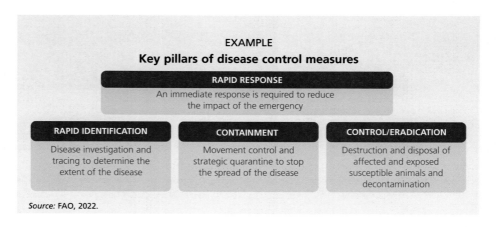

EXAMPLE
Key pillars of disease control measures

RAPID RESPONSE
An immediate response is required to reduce the impact of the emergency

RAPID IDENTIFICATION	**CONTAINMENT**	**CONTROL/ERADICATION**
Disease investigation and tracing to determine the extent of the disease	Movement control and strategic quarantine to stop the spread of the disease	Destruction and disposal of affected and exposed susceptible animals and decontamination

Source: FAO, 2022.

3.3.3.1 Rapid identification

The first 72 hours of an animal health emergency is the most critical time to detect, contain and control the disease in animals.

Therefore, one of the initial objectives should be to locate and quickly confirm the presence of the specific disease pathogen, and the animals affected or at risk. Disease investigation and tracing efforts should target areas and populations at risk in the primary control zone and include activities such as the trace-forward and trace-back of infected animals, exposed animals, contaminated animal products and fomites, to determine the source of infection and stop its spread

To ensure that an appropriate response is implemented as early as possible, surveillance, clinical observation, diagnostic testing and tracing activities are important to establish the extent of infection of premises and zones and to determine the number of affected locations and the risk factors.

3.3.3.2 Containment

The purpose of containment is to control the movement of animals, animal products, vehicles and people into, within or out of the identified zones. Typical containment measures include zoning, biosecurity, quarantine and movement control.

a. Zoning

The definition and the establishment of zones should follow the WOAH Terrestrial Animal Health Code, Chapter 4.4. Zone classifications and a brief description are outlined in the Table 3.

The main factors to be considered when determining zones are the type of disease, the potential to spread, and the geographic features, as well as the commodities and business flows or movements in the areas that are affected.

A typical zoning strategy, as shown in Figure 10, would include two or more concentric zones that extend a prescribed distance from the infected premises.

TABLE 3
Example – WOAH Terrestrial Animal Health Code – Zoning and Compartmentalization

Infected zone	An infected zone is the area in which an infection or infestation has been confirmed. And will normally include the area immediately surrounding infected premises.
Containment zone	A zone which includes all suspected or confirmed cases that are epidemiologically linked and where movement control, biosecurity and sanitary measures are applied to prevent the spread of, and to eradicate, the infection or infestation.
Protection zone	A zone where specific biosecurity and sanitary measures are implemented to prevent the entry of a pathogenic agent into a free country or zone from a neighbouring country or zone of a different animal health status.
Free zone	A zone in which the absence of the disease under consideration or infestation in an animal population has been demonstrated.

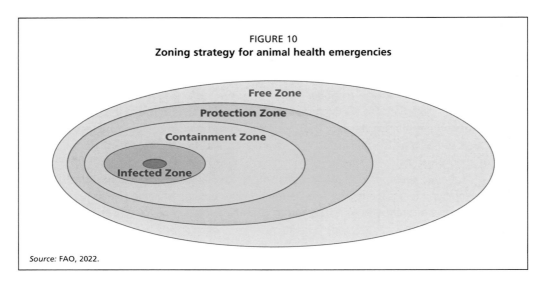

FIGURE 10
Zoning strategy for animal health emergencies

Source: FAO, 2022.

The classification of premises following the first investigations and the risk assessment facilitates the disease investigation and tracing and ensures the adequate implementation of disease control measures. All premises' classifications are subject to change as the animal health emergency evolves. These changes may result from a modification in the case definition, investigations or the successful completion of operational activities.

It should be noted, that while the principles of zoning described in the WOAH Terrestrial Animal Health Code are applied, some countries may assign different names to each zone. An example of how zoning is applied in the United States of America for an FMD outbreak is shown in the following example.

FIGURE 11
Example – zones and areas used in the United States of America for an FMD outbreak

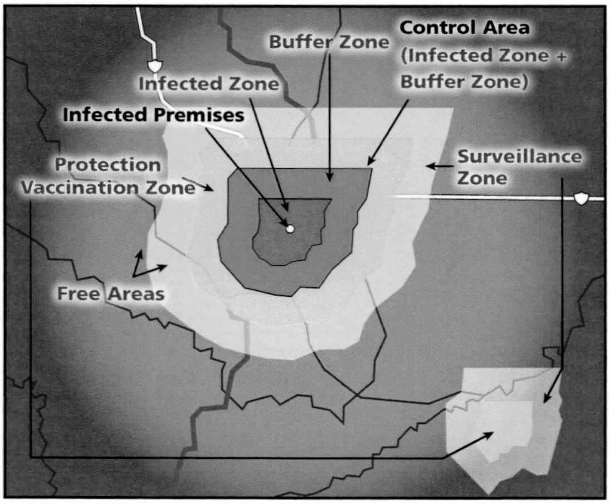

Source: USDA. 2020. Foreign Animal Disease Preparedness and Response Plan (FAD PReP)— Foot-and-Mouth Disease (FMD) Response Plan: The Red Book https://www.aphis.usda.gov/animal_health/emergency_management/downloads/fmd_responseplan.pdf

b. Biosecurity

The biosecurity measures designed to prevent disease spread from infected premises are commonly referred to as biocontainment measures. This differs from bio-exclusion measures, which are designed to prevent the introduction of the disease agent into a naive population. Both measures must be followed during the movement of personnel and materials in and out the protection, containment and infected zones during an animal health emergency. The implementation of biosecurity measures is dependent upon the results of active surveillance being conducted within infected, containment, protection and free zones, and the standing of the zone according to its vaccination, depopulation and disposal status.

In any case, a balance must be maintained between implementing response activities and ensuring that the disease is not being spread. Field responders should follow appropriate SOPs when conducting and implementing biosecurity measures.[1] If necessary, CFPHS may need to be conducted.

c. Quarantine and movement controls

Quarantine and movement control activities are fundamental to rapidly contain an animal health emergency. The veterinary services, based on the existing legal framework, should issue and present quarantine orders to the owner of infected premises as soon as possible, either during the alert phase (for the initial detections) and during the emergency phase as new premises are identified. This will allow field teams to secure the premises and establish biosecurity measures before the implementation of disease control operations. SOPs for quarantine and movement controls should be developed in peacetime.

The establishment of check points help to enforce movement controls and contribute to stopping high-risk movements, which in turn reduces the potential for further disease spread. This may require the involvement of transport authorities, law enforcement, local authorities, industry and community.

Movement control allowances must be made for the humane care of animals (such as feed delivery) and to ensure the continuity of business from non-affected areas. This can be achieved through the issuance of permits to allow movements during an animal health emergency.

USDA's Foreign Animal Disease Preparedness and Response Plan (FAD PReP) Manual, Permitted Movement provides information and guidelines for all stakeholders (USDA, 2017).

3.3.3.3 Control and eradication

The United States of America identifies five strategies (USDA, 2016) for the control and eradication of animal health emergencies in domestic livestock that are not mutually exclusive. Examples of how these strategies can be applied and combined are outlined in the following table.

---

[1] For more information about biocontainment and bio-exclusion biosecurity measures and protocols go to: https://www.cfsph.iastate.edu/pdf/fad-prep-nahems-guidelines-biosecurity

TABLE 4
Example – Emergency response strategies – Foot-and-Mouth disease

Strategy or strategies	Definition of strategy
Stamping-out (no emergency vaccination)	Depopulation of clinically affected and in-contact susceptible animals
Stamping-out modified with emergency vaccination to slaughter	Depopulation of clinically affected and in-contact susceptible animals and vaccination of at-risk animals, with subsequent slaughter of vaccinated animals
Stamping-out modified with emergency vaccination to live	Depopulation of clinically affected and in-contact susceptible animals and vaccination of at-risk animals, without subsequent slaughter of vaccinated animals
Stamping-out modified with emergency vaccination to slaughter and emergency vaccination to live	Combination of emergency vaccination to slaughter and emergency vaccination to live
Vaccination to live (without Stamping-Out)	Vaccination used without depopulation of infected animal or subsequent slaughter of vaccinated animals
No direct action (no stamping-out and no vaccination)	The disease would take its course in the affected population. Measures may be implemented to control the spread (quarantine and movements control, biosecurity measures, cleaning and disinfection measures)

In order to execute the appropriate response strategies, critical activities and tools must be implemented. This includes all options for euthanasia and mass depopulation of the affected and exposed animals after determining the value of the herd or flock. The activities should be provided to the affected animals as safely, quickly and efficiently as possible, and should fully consider animal welfare.

An animal welfare assessment should be included in the initial rapid assessment and used to adjust response measures. Field teams should be able to manage animal welfare issues that arise from the disease or the implementation of disease control measures. This includes the supply of feed and water, humane transportation, euthanasia or culling, and disposal. All the personnel involved in the emergency operations should be trained to comply with the existing animal welfare regulations in emergencies. The WOAH has established guidelines for disaster and risk management in relation to animal welfare (WOAH, 2016).

Proper disposal of animal carcasses and materials, such as bedding, manure and litter, is important for preventing or mitigating pathogen spread and containing, controlling and eradicating the animal disease. The appropriate decontamination methods should be selected based on the characteristics of the disease agent, premises, temperature and other factors. Virus elimination should be completed in the most cost-effective manner possible (Miller et al., 2020).

Depending on the national legal framework and the disease concerned, emergency vaccination may be carried out instead of depopulation or culling. This decision will depend on a number of factors such as:

- economics;
- vaccine suitability;
- nature of the farm operation;
- species involved;
- extent and projected duration of the outbreak;
- resources available (financial or other alternatives); and
- public acceptance.

3.3.4 Reporting and notification obligations

3.3.4.1 Reporting

During the emergency phase, prompt reporting of suspected cases to veterinary services is key for an efficient response. For the purpose of enhancing reporting, veterinary services must through the chain of command and the legal provision achieve the following:

- maintain the link with field veterinary practitioners to ensure full reporting of the suspected cases; and
- ensure that clinical signs of diseases and reporting procedures developed in peacetime are publicized in the appropriate media and provided to producers' associations.

To enable effective maintenance of situational awareness during the emergency phase, responders must collaborate and provide regular and timely reports of the progress of their assigned tasks and any emerging developments. Specific requirements for reporting, such as timing and frequency, must be included in action plans for all responders and management, and situation reports should be based on the situation during the reporting period and not delayed pending completion of other tasks or events.

Briefings and debriefings have been described in Chapter 2. During the emergency phase they should provide information on:

- the aim and operational objectives of the response;
- current strategies being applied to control the disease;
- relevant safety and welfare issues;
- the role of the different stakeholders in the operation; and
- reporting relationships and requirements.

3.3.4.2 Situation reports

Situation reports (SitReps) provide information on the progress of the emergency operations and are disseminated at all levels of the emergency response structures. At least one SitRep should be drafted and disseminated in each operational period and used to inform the relevant incident action plans. An example of a SitRep is provided in Annex 1.

3.3.4.3 Notification

If applicable and following the protocol developed in peacetime for the rapid and transparent reporting of suspected cases, the veterinary national authorities must notify international and regional organizations and neighbouring countries. Cross-border meetings may occur based on the risk of disease spread to neighbouring countries. For the purposes of the Terrestrial Animal Health Code, members of the WOAH have an obligation to immediately (i.e. within 24-hours) report occurrences of notifiable diseases to international organizations, such as WOAH's WAHIS and FAO's Emergency Prevention System for Transboundary Animal and Plant Pests and Diseases (EMPRES), for any of the following:

- first occurrence of a listed disease in a country, a zone or a compartment;
- recurrence of an eradicated listed disease in a country, a zone or a compartment following the final report that declared the event ended;
- first occurrence of a new strain of a pathogenic agent of a listed disease in a country, a zone or a compartment;

- recurrence of an eradicated strain of a pathogenic agent or a listed disease in a country, a zone or a compartment following the final report that declared the event ended;
- a sudden and unexpected change in the distribution or increase in incidence or virulence of, or morbidity caused by the pathogenic agent of a listed disease present within a country, a zone or a compartment; and
- occurrence of a listed disease in an unusual host species.

In addition to this:

- Weekly reports are required subsequent to a notification, to provide further information on the evolution of the event which justified the notification. These reports should continue until the listed disease has been eradicated or the situation has become sufficiently stable to the extent that six-monthly reporting (see below) will satisfy the obligation of the Member Country. For each event notified, a final report should be submitted.
- Six-monthly reports are required on the absence or presence and evolution of listed diseases and information of epidemiological significance to other Member Countries (WOAH, 2021).
- Annual reports concerning any other information of significance to other Member Countries (WOAH, 2021).

At the regional level, there may also be obligations to notify regional organizations such as the African Union Inter-African Bureau for Animal Resources, the European Union, and the Association of Southeast Asian Nations.

Early international reporting and notification will enable early access to international assistance and coordination of response, if required.

3.3.5 Ensuring security and safety of response personnel

Many animal health emergencies are zoonotic and can infect responders conducting operations during an animal health emergency. It is important to take this into account when developing specific control and biosecurity measures. A hierarchy of controls needs to be established to determine how to implement feasible and effective control solutions to protect personnel (CDC, 2017).

Response personnel at all levels should be trained on the appropriate measures to minimize the risk of contamination, and personnel engaged in eradication activities, such as tracing and surveillance, and decontamination, may also be required to be vaccinated for relevant diseases or take prophylactic medicines.

The United States Foreign Animal Disease Preparedness and Response Plan (FAD PReP)/ National Animal Health Emergency Management System (NAHEMS) Guidelines provide guidance on health and safety for responders and outlines the dangers that they might encounter while deployed during an emergency (USDA, 2018).

3.3.6 Monitoring and evaluation during response

The implementation of the monitoring and evaluation activities allows the emergency management system to review how the intervention is progressing, recognize risks and challenges as they develop, and adjust the response. This requires the design of an appropriate monitoring and evaluation system for the management of emergency operations and the identification of the appropriate objectives and indicators. The monitoring of the key progress indicators of the response and their analysis allows for continuous adjustment and adaptation of the control measures (GEMP, 2021). The following table depicts some examples of indicators for monitoring the management of the emergency operations.

EXAMPLE

Indicators for the monitoring operations during an animal health emergency

- ratio of positive to negative reports;
- days from report received to end of culling;
- days from end of culling to end of disposal;
- days from end of disposal to end of cleansing and disinfection;
- numbers of cases, e.g. outbreak locations, during a given period compared with the number of cases in the previous period;
- number of vaccinated animals; and
- number of checkpoints.

Common tools for monitoring and evaluation include debriefings (during and at the conclusion of an animal health emergency) and after-action reviews.

Debriefings should be conducted regularly during the emergency phase of an animal health emergency. This may occur immediately following a task, at the end of the day, following the completion of a group of operational activities and at the conclusion of the emergency phase.

An after-action review should be conducted as soon as possible after the emergency phase has finished and staff have had time to reflect on what has happened.

The FAO Guide to conducting after-action reviews for animal health emergencies[2] provides guidance on planning, preparing for, conducting and reporting on after-action reviews.

3.3.7 Stand down and demobilization
3.3.7.1 Stand down
The stand down begins when:
- the investigation and alert phase fail to confirm the presence of an emergency animal disease;
- the emergency animal disease has been eradicated or controlled; or
- the eradication procedures have failed or are not considered feasible, cost-effective or beneficial and the disease is declared to be established.

---

[2] At the time of writing the FAO Guide to conducting After Action Reviews was being finalised and should be published before the finalisation of this document

Where disease containment, control and eradication operations cease, disease surveillance should continue to enable early detection of any disease recurrence and to demonstrate disease freedom.

The following activities can also take place during the stand down process:

- recover, decommission and dispose of stores and equipment;
- perform a post-incident inventory and replenish/refurbish equipment as needed;
- coordinate documentation (gathering and archiving all documents regarding the incident, including costs and decision making);
- arrange appropriate archiving of all records;
- conduct post incident review and debriefing; and
- help the community to manage expectations by continued communication and disease education campaigns. This will also enhance awareness about the disease status and encourage disease reporting of a potential resurgence or new cases.

3.3.7.2 Demobilization of the emergency response systems

Resources needed for the response should be demobilized as soon as they are no longer needed to manage operations during an animal health emergency. The process for returning them to their day-to-day function should be expedited. Some organizations may be demobilized while other elements are still operational. After de-activation of the emergency response systems, demobilization can follow the process below:

- establish demobilization priorities based on the specific incident;
- define the organization and confirmation of assignment of responsibilities. The on-scene incident commander should approve the release or demobilization of response resources prior to initializing the process;
- disseminate public, stakeholder, and media announcements to provide situational awareness.

3.3.8 Initial recovery actions

Recovery measures for an animal health emergency should be implemented during the emergency phase and continue through the reconstruction phase, as described in the GEMP manual. The implementation of recovery measures is typically longer and can be more expensive than the operational activities undertaken during the emergency phase. The implementation of recovery measures should aim at assisting affected persons, communities and sectors return to normal. At a minimum, the following recovery measures should be implemented at the earliest opportunity:

- provide advice to affected livestock owners about available financial assistance, rebuilding, re-equipping, re-stocking and other agricultural advisory services;
- maintain surveillance activities to establish 'proof of freedom' if appropriate;
- arrange for monitoring of waste disposal sites as appropriate; and
- make arrangements for monitoring of community recovery in accordance with the scale and impact of the animal health emergency.

Further guidance on recovery measures is available in the GEMP manual, while the LEGS handbook provides helpful decision-support tools as part of the recovery process.

Annexes

Annex I
Templates and samples

This annex includes checklists, templates and examples that have been mentioned throughout this manual and can be used to assist in the management of operations during an animal health emergency. These include:

CHECKLISTS
- emergency operations centre – equipment and supplies
- SMEAQ-S briefing format

TEMPLATES
- emergency operations centre assignment list
- incident action plan
- funding requirements form

EXAMPLES
- declaration of an animal health emergency
- standard operating procedure

CHECKLIST – Emergency operations centre – Equipment and supplies

This list is not exhaustive, but contains common items frequently used within an emergency operations centre.

Item	
Computers or laptops with peripherals (keyboard, mouse, monitors, etc.)	❏
Wi-Fi service (if available) or mobile hotspots	❏
Televisions (as many as needed, both to display local networks or maps, info, etc.)	❏
Telephones (landline or mobile) for all staff at the unit level and above	❏
Printers (colour, black & white, and plotter for wall maps) and supplies	❏
White boards or chalk boards and markers/chalk	❏
General office supplies (pens, pencils, notebooks, sticky notes, staplers, paper clips, folders, binders, sheet protectors, etc.)	❏
Microphone and a PA system (for briefings)	❏
Generator(s) and fuel (if power supply is non-existent or unreliable)	❏
First aid kit for staff	❏
Food service items	❏
Catering supplies and cooking equipment	❏
Refrigerator or freezer	❏
Cleaning and toilet facility supplies	❏
Bottled water or water purification equipment	❏

CHECKLIST – SMEACS-Q briefing approach

Briefing outputs using the **SMEACS-Q** approach will address the following topics:

- **S = Situation**. What are we facing, what is likely to happen, what are the issues of which we need to be aware? Describes an overview of the incident, a summary of resources already deployed, current expected weather, and known risks.

- **M = Mission**. What outcome is to be achieved, and, most importantly, why? Given in terms of specific objectives for the response. (Field teams at the action interface will translate this into "How" to use SOPs, policies, dynamic risk assessments and safety protocols).

- **E = Execution**. The incident manager and incident management team strategies and tactics, using the incident action plan (IAP). Who is being activated, where and when? Access to the incident, contingency plans, immediate tasks after briefing. (Field team actions are decided by the team leader according to the SOPs, policies and the specific objectives above.)

- **A = Administration/logistics**. Resources to hand out being sourced or acquired. Personnel deployment, supply and support to the mission.

- **C = Command and communication**. Who do you report to, what is the chain of command, what is the communications plan, and what are the contact numbers and radio channels?

- **S = Safety**. Identify known or likely hazards and risks relevant to the operational period. Personal protective equipment (PPE), weather, hydration and first aid.

- **Q = Questions**. These occur at the very end of the briefing, providing an opportunity for clarification, to receive additional details to enable full understanding. There is no debate.

TEMPLATE – Emergency operations centre assignment List

INCIDENT COMMAND			
Position	**Name**	**Agency/Office**	**Contact details**

OPERATIONS			
Position	**Name**	**Agency/Office**	**Contact details**

PLANNING			
Position	**Name**	**Agency/Office**	**Contact details**

LOGISTICS			
Position	**Name**	**Agency/Office**	**Contact details**

FINANCE/ADMIN			
Position	**Name**	**Agency/Office**	**Contact details**

Source: Office of U.S. Foreign Disaster Assistance (OFDA).

TEMPLATE – Incident action plan

EOC NAME: ....

INCIDENT/EVENT NAME: ....

OPERATIONAL PERIOD:

From (Date and Time): ....

To (Date and Time): ....

PREPARED BY:	**APPROVED BY:** ....
Date/Time Prepared:	Date/Time Approved: ....

SENIOR LEADERSHIP PRIORITIES

1. ....

2. ....

3. ....

OVERALL INCIDENT OBJECTIVES	EOC OBJECTIVES
1.	**1.** ....
2.	**2.** ....
3.	**3.** ....
4.	**4.** ....

ATTACHMENTS (CHECK ALL THAT APPLY)	
☐ **Organization assignment list**	☐ **Others** ....
☐ **Meeting schedules**	☐ **Others** ....
☐ **Directory**	☐ **Others** ....
☐ **Situation map**	☐ **Others** ....

MANAGEMENT FUNCTION TASKS	Assigned to:
....	....

OPERATIONS FUNCTION TASKS	Assigned to:
....	....

PLANNING FUNCTION TASKS	Assigned to:
....	....

LOGISTICS FUNCTION TASKS	Assigned to:
....	....

FINANCE/ADMIN FUNCTION TASKS	Assigned to:
....	....

TEMPLATE – Emergency funding requirements form

FUNDING REQUIREMENTS:

OPERATING PERIOD   from: ....   to: ....

ITEM	QUANTITY	COST (ESTIMATE)
ACCOMMODATION		
Hotel/other		
Catering		
Communications		
Telecommunications		
Other (e.g. radio)		
Advertisements		
CONTRACTORS/SERVICE PROVIDERS		
Machinery hire, etc..		
CONTROL CENTRE		
Telecommunications		
Electricity		
Security		
Venue hire		
Display boards,		
Whiteboards		
Maps		
EMERGENCY SUPPLIES		
Disinfectant		
Equipment for animal destruction		
Protective clothing		
Plastic boxes e.g. 50l (for transport of equipment), also as portable footbath		
Plastic scrubbing brushes		
Tarpaulins, ground sheets		
Tent (portable 'changing' room)		
Water containers (e.g. 20l)		
Buckets		
Plastic backpacks (e.g. 15l) for decontamination		
Transport		
Airfares		
Bus fare		
Vehicle hire		
Vessel costs		
Other		

Source: UNDRR. 2009. Agricultural Emergency Response Plan Template. https://www.preventionweb.net/files/27076_cookislandsanimalhealthemergencyres.doc

EXAMPLE – Declaration of animal health emergency

Ontario
Executive Council
Conseil des ministres

Order in Council
Décret

On the recommendation of the undersigned, the Lieutenant Governor, by and with the advice and concurrence of the Executive Council, orders that:

Sur la recommandation du soussigné, le lieutenant-gouverneur, sur l'avis et avec le consentement du Conseil des ministres, décrète ce qui suit :

On the recommendation of the undersigned, the Lieutenant Governor, by and with the advice and concurrence of the Executive Council, orders that:

PURSUANT to Subsection 6(1) of the *Emergency Management Act* R.S.O. 1990 c.E9, as amended, all ministers are responsible for the formulation of emergency plans in respect of any emergency that affects the continuity of operations and services in their respective ministries.

In addition to the above, the following ministers are responsible for the formulation of emergency plans in respect of the type of emergency assigned.

Minister	Type of Emergency
Agriculture, Food and Rural Affairs	Farm animal disease; food contamination; agricultural plant disease and pest infestation
Attorney General	Any emergency related to the administration of justice including the operation of the courts; and provision of legal services to government in any emergency
Community and Social Services	Any emergency that requires emergency shelter, clothing and food; victim registration and inquiry services; personal services

O.C./Décret 1492/2005

EXAMPLE – Standard operating procedures

NATIONALLY AGREED STANDARD OPERATING PROCEDURE (NASOP)

Title: **Collecting emergency animal disease samples for laboratory testing**

Version: **1.0**

Prepared by: **Subcommittee on Emergency Animal Disease**

Approved by: **Animal Health Committee**

Revision history:

Version	Date of approval	Comments
1.0	02/05/11	Approved by AHC

NASOPs support national consistency and provide guidance to response personnel undertaking operational tasks.

1. Purpose
- This document guides the emergency animal disease (EAD) sample collection procedures to be used by sampling teams, operating from operations centres, who collect EAD samples for laboratory testing. This NASOP should be supported by jurisdictional SOPs and disease specific control strategy manuals.

NOTE: Disease-specific sample collection protocols may need to be developed by operations centres during the response in collaboration with the designated laboratory.

2. Application/Scope
- This NASOP should be used by relevant personnel in operations centres and laboratories to guide the development of disease specific sample collection protocols.
- The final disease specific sample collection protocol will include the choice of samples, collection and transport, priority of testing and release of test results.
- Separate to this NASOP, an operations centre should ensure that:
  - samples are collected, packaged and transported correctly to ensure 'chain of custody with samples' requirements applicable to the jurisdiction and the disease response
  - samples are packaged according to appropriate transport regulations
  - State Coordination Centre (SCC) has chosen a testing laboratory for the response
  - the laboratory is provided with the best samples possible
  - relevant transport regulations are complied with.
- The appropriate AUSVETPLAN disease strategy outlines the specific diagnostic samples and their collection for isolation of specific suspect causative agents.
- The appropriate AUSVETPLAN disease strategy provides guidance on the samples required for surveillance and proof of freedom and the agreed level of collection for the outbreak.

3. Resources/equipment
- clear instruction on the number and status of premises to be visited for each trip out from the operations centre
- sampling protocol(s) e.g. tissues to be collected, number of samples per herd/flock
- appropriate sampling equipment (consider sample kit checklist at Appendix A)
- owner/manager and livestock details including numbers and locations for each property

Source: Animal Health Australia (AHA), 2011.

Annex II
Guidelines for establishing an emergency operations centre

This annex provides guidance on how to set up, organize, and operate an emergency operations centre for an animal health emergency. It is not meant to be exhaustive, but rather a resource to help guide decisions to establish, staff, and put into effect an emergency operations centre to manage operations during an animal health emergency.

This annex has been informed by the Office of U.S. Foreign Disaster Assistance, EOC Quick Start Guide and AUSVETPLAN Control Centre Management Manual and provides guidance on:

- considerations for selecting a site for an emergency operations centre;
- layout of an emergency operations centre;
- initial actions to establish an emergency operations centre;
- emergency operations centre positions and staffing; and
- emergency operations centre key position checklists.

1. CONSIDERATIONS FOR SELECTING A SITE FOR AN EMERGENCY OPERATIONS CENTRE

The following provides guidance on identifying and establishing a suitable site for an emergency operations centre for the management of operations during an animal health emergency.

Primary considerations

Once established, the emergency operations centre will be a focal point for the management of operations during an animal health emergency. As it may need to be maintained for an extended period of time, it is important that every effort is given to ensure that it is set-up from the outset, to accommodate the ongoing needs of the animal health emergency.

The emergency operations centre should be identified as early as possible and be located:

- within reasonable travel time from where operational activities are being conducted;
- in a low-risk area, ideally a safe distance from infected premises; and
- near or accessible to accommodation, transport, service providers (catering, cleaning etc).

Options for establishing an emergency operations centre include:

- existing government buildings;
- another agency's emergency operations centre;
- community halls; and
- disused commercial premises, such as a factory, warehouse or car sale yard.

General considerations

Using the criteria identified above, selecting a site for an emergency operations centre should also be mindful of the impact that it may have on surrounding communities and businesses, as there is likely to be a lot of movement to and from the emergency operations centre that could disrupt normal community activities.

The size of the emergency operations centre needs to be sufficient to accommodate the number of people who are likely to be working there, visiting and transiting through the emergency operations centre, where allowances may need to be made for work areas, catering, briefings and potential storage of operational equipment and consumables.

Considerations for layout of the emergency operations centre are covered in the next section of this annex.

2. LAYOUT OF AN EMERGENCY OPERATIONS CENTRE

A number of factors will determine the size and layout of an emergency operations centre. In many cases, size may be limited by the spaces are available for use, and staff may need to get creative with layouts to improve functionality and flow.

General considerations for the size of an emergency operations centre should include:

- sufficient space for staff assigned to work within the emergency operations centre;
- space for briefings, meetings, media, visitors;
- space for eating and resting;
- designated parking area;
- sufficient space for equipment requirements of the emergency operations centre; and
- space for operational activities, including storage of equipment, consumable items and hazardous materials, if required.

Once a location has been determined and staff have a sense of the overall size of the space available, layout can be considered. The layout will be dependent upon a number of factors, including:

- Number of personnel and site constraints.
- All emergency operations centres should have a central operations room that serves as the "nerve centre." The operations room is a large area designed to facilitate coordination, communications, resource management and decision-making.
- Other rooms to support operational activities may include:
  - dedicated offices or workspaces for senior staff;
  - meeting/conference room/s for briefings, debriefings, meetings, trainings and other uses;
  - a secure space for media/press briefings, and visitors outside the primary operations room;
  - space for communications equipment (radio room); and
  - separate space for personnel support, such as a kitchen, pantry, rest area, comfort room and showers.

An emergency operations centre may not need to contain all the above. In some cases, the most appropriate space may be a single room with a conference table. The following are suggestions for the layout of an emergency operations centre.

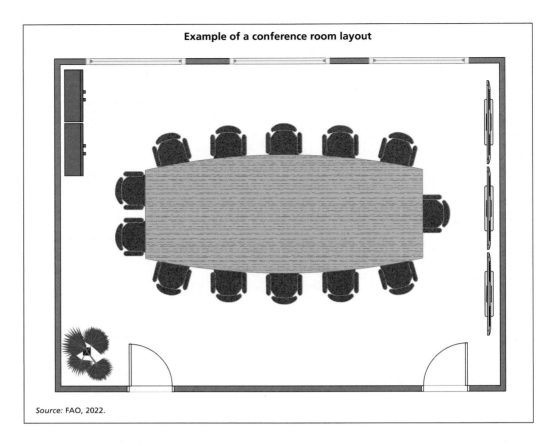

Example of a conference room layout

Source: FAO, 2022.

Conference Room

A conference room layout gathers staff around a single table and enables close collaboration during operations. This layout is typically used in smaller spaces and takes very little additional equipment to set up.

This layout is useful for senior staff to come together to meet, discuss and agree on operational activities.

When this approach is used, space will also be required for staff and workstations required to undertake the work necessary to manage operations during an animal health emergency.

The National Disaster Risk Reduction and Management Council (NDRRMC), Operation Center – Manila, Philippines

Source: NDRRMC Facebook page - www.facebook.com/NDRRMC

Situation Room

This layout is based on a dedicated space within the emergency operations centre where staff are seated in rows or semi-circles facing large visual displays. The situation room (sometimes referred to as the operations room) visually displays current information and is heavily dependent on technology. Staff communicate primarily through incident management software.

This layout works well for technical tasks but may limit collaboration and interaction among staff and is among the costliest options.

A less costly option may be to find a space in an actual university lecture hall; the layout is similar, and much of the equipment needed could be procured elsewhere and installed for the duration of the response.

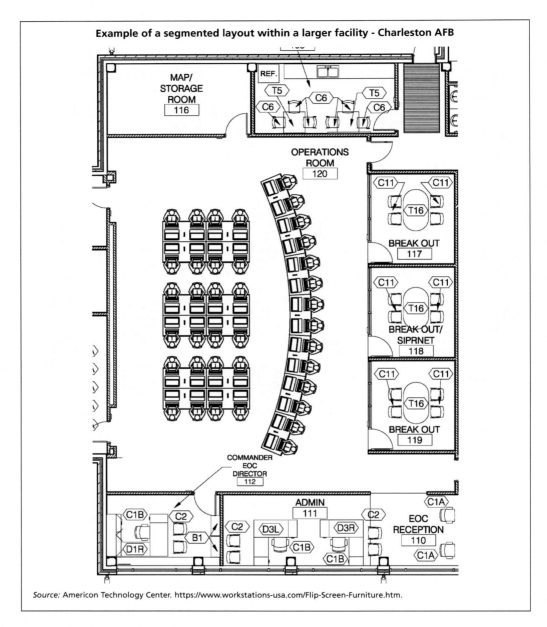

Example of a segmented layout within a larger facility - Charleston AFB

Source: Americon Technology Center. https://www.workstations-usa.com/Flip-Screen-Furniture.htm.

Segmented Layout

In this layout, staff are separated into function-specific groups (such as clusters, agencies, or emergency operations centre management functions) that emphasize close collaboration and flexible interaction between functions and groups. It provides a high level of autonomy for each function/group.

Coordination is accomplished by moving among the various groups. This is a useful layout when an emergency operations centre has a large number of staff and an adequate amount of space to accommodate everyone.

3. INITIAL ACTIONS TO ESTABLISH AN EMERGENCY OPERATIONS CENTRE

The logistics section is normally responsible for locating and establishing a suitable facility to use as the emergency operations centre for managing operations during an animal health emergency. This may include the appointment of a dedicated facilities manager, within the logistics section, who is also responsible for the efficient and effective running of the emergency operations centre.

Tasks that may need to be undertaken by the facilities manager include:

- Meet with the Incident Commander and Section Chiefs to determine: current situation and priorities for the emergency operations centre; anticipated trajectory of the response; current resources and anticipated resource requirements; and communication requirements between the emergency operations centre and field personnel.
- Identify staff required to establish and set-up the emergency operations centre.
- Acquire equipment and supplies required for the emergency operations centre.
- Identify and implement safety and security requirements to ensure safety and security of the emergency operations centre and its staff.
- Develop and disseminate an organisation chart and staffing plan identifying allocated areas for each section within the emergency operations centre.

As staff are assigned to the emergency operations centre and begin arriving, staff from the logistics section will provide briefings on the situation, an overview of the emergency operations centre and assigned workspace, and expectations.

4. EMERGENCY OPERATIONS CENTRE POSITIONS AND STAFFING

The organizational structure of the emergency operations centre should mirror the incident command system organisational structure, shown in the following diagram.

The EOC **incident commander** oversees the staff and activities within the operations

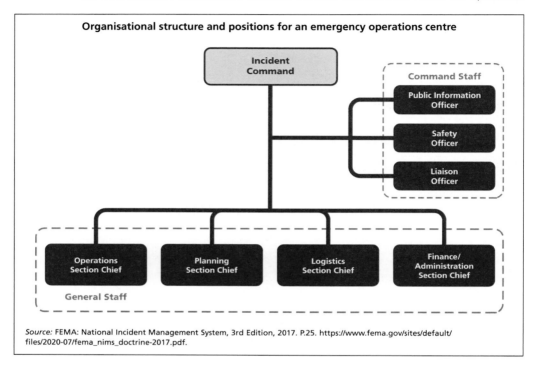

Organisational structure and positions for an emergency operations centre

Source: FEMA: National Incident Management System, 3rd Edition, 2017. P.25. https://www.fema.gov/sites/default/files/2020-07/fema_nims_doctrine-2017.pdf.

centre and reports up the chain of command.

The incident commander determines the scope of the operational response to the animal health emergency and the appropriate staffing levels within the emergency operations centre. In this position, the incident commander is also expected to coordinate externally from the emergency operations centre with higher level authorities and to those parties apprised of emergency operations centre's activities.

There are a few positions that work tangentially to the incident command section and report directly to the incident commander. Recommended positions in the command section include:

- **public affairs/Information officer** to coordinate all media releases and all press briefings if needed;
- **safety/security officer**, whose main responsibilities are to ensure the EOC facility is in good condition and poses no threat to the team, to be aware of any external threat to the team, and to make sure that all operational activities that occur within the facility are safe; and
- **liaison officer** responsible for assisting the incident commander by serving as point of contact for representatives from other response organizations.

The incident commander oversees four sections within the emergency operations centre. The sections are **operations, planning, logistics,** and **finance and administration**:

- **The operations section**, led by the **operations section chief**, coordinates response specific requirements and needs. The specific functions that are stood up in the operations section are specific to the individual needs of the animal health emergency. For example (destruction, disposal and decontamination unit, infected premises operations unit, movement control unit, veterinary investigations unit).

- The **planning** section, led by the **planning section chief**, produces different informational products that track resources, as well as creates situational reports. The planning section also collects, collates, analyses, and disseminates information throughout the sections within the emergency operations section. Examples of units with the planning section include the situation unit, resources unit, and the documentation unit.

- The **logistics** section, led by the **logistics section chief**, maintains the emergency operations centre facility and the equipment used within it as well as transporting goods, food and medical supplies for all personnel assigned to the emergency operations centre. Some examples of logistics section units include the staffing unit, equipment and supplies unit, facilities unit, and the catering and accommodation unit.

- The **finance and administration** section, led by the **finance/admin section chief**, manages all financial aspects of the emergency operations centre and pays all expenses incurred by the emergency operations centre. Examples of finance/admin section units include accounting/audit (cost recovery) unit, and the fiscal (procurement) unit.

5. EMERGENCY OPERATIONS CENTRE KEY POSITION CHECKLISTS
Emergency operations centre – Common responsibilities checklist
READ ENTIRE CHECKLIST BEFORE TAKING ANY ACTION!

❏ **JOB ASSIGNMENT** Receive assignment from your department, including:
  ○ job assignment;
  ○ reporting location;
  ○ reporting time;
  ○ travel instructions;
  ○ any special communications instructions, e.g. phone numbers, travel frequency; and
  ○ order and request numbers.

❏ **CHECK-IN** upon arrival at the emergency operations centre or assigned work locations.

❏ **INITIAL BRIEFING:** Receive briefing from immediate supervisor.

❏ **WORK MATERIALS:** Acquire work materials.

❏ **HEALTH AND SAFETY:** Conduct all tasks in a manner that ensures your safety and welfare and that of your co-workers.

❏ **SUPERVISION:** Organize and brief subordinates if in supervisory role.

❏ **COMMUNICATION:** Know the assigned telephone or mobile numbers and radio frequency(ies) (if necessary) for your area of responsibility and ensure that communication equipment is operating properly. Use clear text and terminology in all radio communications.

❏ **DOCUMENTATION:** Complete forms and reports required of the assigned position and send through the chain of command to the documentation unit.

❏ **DEMOBILIZATION:** Respond to demobilization orders and brief subordinates regarding demobilization.

Emergency operations centre (EOC) incident commander position checklist

READ ENTIRE CHECKLIST BEFORE TAKING ANY ACTION!

❑ Review common responsibilities – All personnel.

❑ Determine appropriate level of activation based on situation as known.

❑ Mobilize appropriate personnel for the initial activation of the EOC.

❑ Establish the appropriate staffing level for the EOC and continuously monitor organizational effectiveness, making appropriate modifications as required.

❑ Ensure that the EOC is properly set up and ready for operations.

❑ Ensure that an EOC organization and staffing chart is posted and completed.

❑ Determine which general staff are needed, assign section chiefs as appropriate and ensure they are staffing their sections as required.

❑ In coordination with section chiefs, set priorities for response efforts.

❑ Ensure that communications with supporting agencies are established and functioning.

❑ Confer with the section chiefs to determine what representation is needed at the EOC from member jurisdictions and other emergency response agencies.

❑ Assign a liaison officer to coordinate outside agency response to the EOC.

❑ In coordination with the public information officer (PIO), conduct news conferences and review media releases for final approval, following the established procedure for information releases and media briefings.

❑ Ensure that the liaison officer is providing and maintaining effective interagency coordination.

❑ Based on current status reports, establish initial objectives for the EOC and operations for the animal health emergency.

❑ In coordination with general staff, prepare management function objectives for the initial action planning meeting.

❑ Convene the initial action planning meeting. Ensure that all section chiefs, management staff, and other key agency representatives are in attendance.

❑ Once the incident action plan is completed by the planning section, review, approve and authorize its implementation.

❑ Conduct periodic briefings with the general staff to ensure objectives are current and appropriate.

❑ Conduct regular briefings for responsible officials and/or designated alternates.

❑ Brief your relief at shift change, ensuring that ongoing activities are identified, and follow-up requirements are known.

❑ Authorize deactivation of sections, branches and units when they are no longer required.

❑ Notify other EOCs (if activated), emergency response agencies and other appropriate organizations of the planned deactivation time.

❑ Ensure that any open actions not yet completed will be handled after deactivation.

❑ Ensure that all required forms or reports are completed prior to deactivation.

❑ Be prepared to provide input to the after-action report

❑ Deactivate the EOC at the designated time, as appropriate.

Public information officer (PIO) position checklist

READ ENTIRE CHECKLIST BEFORE TAKING ANY ACTION!

❏ Review common responsibilities – All personnel.

❏ Serve as the coordination point for all media releases for the EOC.

❏ Represent the EOC as the lead public information officer.

❏ Ensure the public within the affected area receives complete, accurate and consistent information about the animal health emergency, public health advisories, relief and assistance programmes and other vital information.

❏ Coordinate the provision of situation information on the relevant government agency websites.

❏ Organize the format for press conferences in coordination with the EOC incident commander.

❏ Obtain policy guidance from the EOC incident commander on media releases.

❏ Keep the EOC incident commander advised of all unusual requests for information and all major critical or unfavourable media comments.

❏ Coordinate with the situation unit and identify method for obtaining and verifying significant information as it develops.

❏ Develop and publish a media briefing schedule, to include location, format, preparation and distribution of handout materials.

❏ Establish a media information centre, if necessary.

❏ Maintain up-to-date status boards and other references at the media information centre.

❏ Interact with other EOC sections, branches, and units to provide and obtain information relative to public information operations.

❏ At the request of the EOC incident commander, prepare media briefings for elected/executive representatives and other government officials, and provide assistance as necessary to facilitate their participation in media briefings and press conferences.

❏ Provide sufficient staffing and telephones for call centre operations to efficiently handle incoming media and public calls.

❏ Monitor broadcast media, using information to develop follow-up news releases and rumour control.

❏ Ensure file copies are maintained of all information releases (for submission to planning section).

❏ Provide copies of all media releases to the EOC incident commander.

❏ Conduct shift change briefings, ensuring that in-progress activities are identified, and follow-up requirements are known.

❏ Prepare final news releases and advise media representatives of points-of-contact for follow-up stories.

Operations section chief position checklist

READ ENTIRE CHECKLIST BEFORE TAKING ANY ACTION!

❏ Review common responsibilities – All personnel.

❏ Ensure the operations function is carried out including coordination of response for all operational functions assigned to the EOC.

❏ Ensure operational objectives and assignments identified in the IAP are carried out effectively.

❏ Establish the appropriate level of organization for the operations section, continuously monitoring its effectiveness and modifying accordingly.

❏ Ensure the planning section is provided with status reports on a regular schedule.

❏ Conduct periodic operations briefings for the EOC incident commander, as required or requested.

❏ Provide overall supervision of the operations section.

❏ Ensure that the operations section is set up properly and that appropriate personnel, equipment and supplies are in place, including maps and status boards.

❏ Meet with planning section chief to obtain a preliminary and regular situation briefings.

❏ Coordinate with the liaison officer regarding the need for agency representatives in the operations section.

❏ Establish radio or cell phone communications with field staff.

❏ Determine activation status of other EOCs and establish communication links with their operations sections.

❏ Based on the situation known or forecasted, determine likely future needs of the operations section.

❏ Identify key issues currently affecting the operations section, meet with section personnel to determine appropriate section objectives for the first operational period.

❏ Review responsibilities of branches in section to develop an operations plan detailing strategies for carrying out operations objectives.

❏ Ensure that all media contacts are referred to the public information section.

❏ Conduct periodic briefings and work to reach consensus among staff on objectives for forth-coming operational periods.

❏ Attend and participate in EOC action planning meetings.

❏ Provide the planning section chief with the operations section's objectives prior to each action planning meeting.

❏ Ensure that the operations section objectives, as defined in the current AP, are being addressed.

❏ Ensure that the units within the operations section coordinate all resource needs through the appropriate contacts in the logistics section.

Planning section chief position checklist

READ ENTIRE POSITION CHECKLIST BEFORE TAKING ANY ACTION!

- ❏ Review common responsibilities checklist.
- ❏ Ensure that the responsibilities of the planning section are carried out, to include:
  - ○ collecting, collating, analysing, and displaying situation information;
  - ○ preparing periodic situation reports;
  - ○ preparing and distributing the EOC incident action plan;
  - ○ facilitating the action planning meeting;
  - ○ conducting advance planning activities and reports;
  - ○ providing technical support services to the various EOC sections and units; and
  - ○ documenting and maintaining files on all EOC activities.
- ❏ Establish the appropriate level of organization for the planning section, continuously monitoring its effectiveness and modifying accordingly.
- ❏ Ensure the early and continued coordination with the planning sections of other EOCs (i.e. local, state or federal EOCs).
- ❏ Exercise overall responsibility for the coordination of unit activities within the section.
- ❏ Keep the EOC incident commander informed of significant issues affecting the planning section.
- ❏ In coordination with the other section chiefs, ensure that status reports and situation reports are used to develop the IAP.
- ❏ Ensure that the planning section is set up properly and appropriate personnel, equipment, and supplies are in place, including maps and status boards.
- ❏ Meet with operations section chief and obtain and review any major incident reports.
- ❏ Identify key issues to be addressed in the EOC action planning process by consulting with section chiefs, including specific objectives to be accomplished during the initial and subsequent operational periods.
- ❏ Keep the EOC incident commander informed of significant events.
- ❏ Ensure that the situation unit is maintaining current information for situation reports.
- ❏ Ensure major incidents reports and branch status reports are completed by the operations section and are accessible by planning section.
- ❏ Ensure a situation report is produced and distributed to all EOC Sections at least once for each operational period.
- ❏ Ensure all status boards and other displays are kept current and that posted information is neat and legible.
- ❏ Facilitate the EOC action planning meetings.
- ❏ Ensure objectives for each section are completed, collected, and posted in preparation for the next action planning meeting.
- ❏ Ensure the EOC IAP is completed and distributed at the start of the next operational period.
- ❏ Work closely with each unit within the planning section to ensure the section objectives as defined in the current EOC IAP are being addressed.
- ❏ Ensure the documentation unit maintains files on all activities related to the event and provides reproduction services for the EOC.

Logistics Section Chief position checklist

READ ENTIRE CHECKLIST BEFORE TAKING ANY ACTION!

❏ Review common responsibilities – all personnel.

❏ Ensure the logistics function is carried out.

❏ Establish the appropriate level of organization within the logistics section, continuously monitoring its effectiveness and modifying accordingly.

❏ Ensure the planning section is provided with status reports on a regular schedule.

❏ Ensure that the logistics section is set up properly and that appropriate personnel, equipment, and supplies are in place.

❏ Meet with planning section chief to obtain a preliminary and subsequent situation briefings.

❏ Confer with the EOC incident commander to ensure that the operations, planning and finance and administration sections are staffed at levels necessary for the response operations.

❏ Provides necessary equipment for the operations section to establish and maintain radio or mobile phone communications with field staff.

❏ Determine activation status of other EOCs and establish communication links with their logistics sections.

❏ Based on the situation known or forecasted, determine likely future needs of the Logistics Section.

❏ Review responsibilities of units in the section to develop a logistics plan detailing strategies for carrying out logistics objectives.

❏ Ensure that all media contacts are referred to the public information section.

❏ Attend and participate in EOC action planning meetings.

❏ Provide the planning section chief with the logistics section's needs prior to each action planning meeting.

❏ Ensure that the logistics section tasks, as defined in the current AP, are being addressed.

❏ Ensure that the units within the operations section coordinate all resource needs through the appropriate contacts in the logistics section.

Finance section chief position checklist

READ ENTIRE CHECKLIST BEFORE TAKING ANY ACTION!

❑ Review common responsibilities – All personnel.

❑ Ensure that all financial records are maintained throughout the event or disaster.

❑ Ensure that all on-duty time is recorded for each person staffing the EOC.

❑ Ensure that all on-duty time sheets are collected from field level supervisors or EOC incident commanders and their staff who are assigned to the response.

❑ Determine a payroll process for response staff if one does not already exist.

❑ Determine purchase order limits for the procurement function.

❑ Ensure that workers' compensation claims, resulting from the response to the event or disaster by employees, are processed within a reasonable time, given the nature of the situation.

❑ Ensure that all travel and expense claims are processed within a reasonable time, given the nature of the situation.

❑ Provide administrative support to the EOC sections as required, in coordination with the personnel unit.

❑ Activate units within the finance section as required and monitor section activities continuously and modify the organization as needed.

❑ Ensure that all recovery documentation is accurately maintained during the response and submitted on the appropriate forms to relevant relief agencies.

❑ Ensure there is coordination with all EOCs for the purpose of gathering and consolidating response cost estimates and other related information.

❑ Meet with the logistics section chief and review financial and administrative support requirements and procedures. Determine the level of purchasing authority.

❑ If there is any indication that the jurisdiction can no longer support the costs of the response and/or recovery, the EOC incident commander needs to be informed immediately.

❑ Ensure that displays associated with the finance section are current and that information is posted in a legible and concise manner.

❑ Participate in all action planning meetings.

❑ Keep the EOC incident commander, general staff, and individual local, state or federal departments engaged in the response aware of the current fiscal situation and other related matters, on an on-going basis.

❑ Ensure the cost recovery unit maintains all financial records throughout the event or disaster.

❑ Ensure sections and units are coding their time correctly in accordance with the specific SOPs for cost tracking.

❑ Ensure that the procurement unit processes purchase orders and develops contracts in a timely manner.

❑ Ensure that the compensation and claims unit processes all claims resulting from the animal health emergency, in a reasonable time frame, given the nature of the situation.

❑ Ensure that the time-keeping unit processes all time sheets and travel/expense claims promptly through the appropriate budget and payroll office.

❑ Ensure that the finance section provides administrative support to other EOC Sections as required.

References

Animal Health Australia. Emergency Animal Disease Response Agreement
https://animalhealthaustralia.com.au/eadra/

Animal Health Australia. 2019. *AUSVETPLAN Control Centre Management Manual, Part 1 and 2*
https://animalhealthaustralia.com.au/ausvetplan/

Animal Health Australia. 2020. Government and Livestock Cost Sharing Deed in Respect of Emergency Animal Disease Response

Animal Health Australia. *Nationally Agreed Standard Operating Procedures.*
https://animalhealthaustralia.com.au/nationally-agreed-standard-operating-procedures/

CDC. Centre for Disease Control and Prevention. 2017. *Hierarchy of Controls Applied to NIOSH TWH,* https://w ww.cdc.gov/niosh/twh/guidelines.html

The Center for Food Security and Public Health. *Just-in-Time Training for Responders.*
https://www.cfsph.iastate.edu/emergency-response/just-in-time-training/

Colombia Ministry of Agriculture. https://www.minagricultura.gov.co/Paginas/fondos_parafiscales.aspx

European Union. COMMISSION DECISION of 28 February 2007 establishing a Community Veterinary Emergency Team to assist the Commission in supporting Member States and third countries in veterinary matters relating to certain animal diseases
https://eur-lex.europa.eu/legal-content/EN/TXT/PDF/?uri=CELEX:32007D0142&from=EN

FAO. 2011. *Good Emergency Management Practice, Standard Operating procedures for HPAI Response* http://www.fao.org/3/a-i2364e.pdf

FAO. 1999. *Manual on Livestock Disease Surveillance and Information Systems.*
https://www.fao.org/3/x3331e/X3331E00.htm

FAO. A guide to compensation schemes for livestock disease control *http://www.fao.org/ag/againfo/resources/documents/compensation_guide/introduction.html*

FAO, WOAH, WHO. 2019. *A Tripartite Guide to Addressing Zoonotic Diseases in Countries*
http://www.fao.org/3/ca2942en/ca2942en.pdf

FAO. 2021. *FAO Strategy for Private Sector Engagement, 2021-2025.* Rome.
http;//doi.org/10.4060/cb3352en

Gary, F., Clauss, M., Bonbon, E. & Myers, L. 2021. *Good emergency management practice: The essentials – A guide to preparing for animal health emergencies. Third edition.* FAO Animal Production and Health Manual No. 25. Rome, FAO. https://doi.org/10.4060/cb3833en

Inter-Agency Standing Committee. 2018. *Standard Operating Procedures, Humanitarian system-wide scale-up activation, Protocol 1: Definition and Procedures* https://interagencystandingcommittee.org/system/files/181113_protocol_1_-_system-wide_scale-up_activation_final.pdf

Miller, L.P., Miknis, R.A. and Flory, G.A. 2020. *Carcass management guidelines – Effective disposal of animal carcasses and contaminated materials on small to medium-sized farms.* FAO Animal production and health Guidelines no. 23. Rome, FAO. https://doi.org/10.4060/cb2464en

World Organisation for Animal Health (WOAH). 2021. *Terrestrial Animal Health Code,* https://www.woah.org/en/what-we-do/standards/codes-and-manuals/terrestrial-code-online-access/

World Organisation for Animal Health (WOAH). 2016. *Guidelines on Disaster Management and Risk Reduction in Relation to Animal Health and Welfare and Veterinary Public Health.* https://www.woah.org/app/uploads/2021/03/disastermanagement-ang.pdf

World Organisation for Animal Health (WOAH). 2019. *Guidelines for Public-Private Partnerships in the veterinary domain.* https://www.woah.org/fileadmin/Home/eng/Media_Center/docs/pdf/PPP/oie_ppp_handbook-20190419_ENint_BD.pdf

World Organisation for Animal Health (WOAH). *Emergency and Resilience* https://www.woah.org/en/what-we-offer/emergency-and-resilience/

World Organisation for Animal Health (WOAH). 2021 .*Terrestrial Animal Health Code. https://www.woah.org/en/what-we-do/standards/codes-and-manuals/terrestrial-code-online-access/*

UNDRR. 2009. *Agricultural Emergency Response Plan Template.* https://www.preventionweb.net/files/27076_cookislandsanimalhealthemergencyres.doc

USDA. 2016. *HPAI Preparedness and Response Plan* https://www.aphis.usda.gov/animal_health/downloads/animal_diseases/ai/hpai-preparedness-and-response-plan-2015.pdf

USDA. 2016. *Foreign Animal Disease Response, Ready Reference Guide – Roles and Coordination* https://www.aphis.usda.gov/animal_health/emergency_management/downloads/fad_prep_rrg_roles_coordination.pdf

USDA. 2018. *The Foreign Animal Disease Preparedness and Response Plan (FAD PReP)/National Animal Health Emergency Management System (NAHEMS) Guidelines* https://www.aphis.usda.gov/animal_health/emergency_management/downloads/nahems_guidelines/fadprep-nahems-guidelines-health-safety.pdf

USDA. 2020. *Foreign Animal Disease Preparedness and Response Plan (FAD PReP)— Foot-and-Mouth Disease (FMD) Response Plan: The Red Book* https://www.aphis.usda.gov/animal_health/emergency_management/downloads/fmd_responseplan.pdf

FAO ANIMAL PRODUCTION AND HEALTH MANUAL

1. Small-scale poultry production, 2004 (En, Fr)
2. Good practices for the meat industry, 2004 (En, Fr, Es, Ar)
3. Preparing for highly pathogenic avian influenza, 2007 (En, Ar, Es^e, Fr^e, Mk^e)
3. Revised version, 2009 (En)
4. Wild bird highly pathogenic avian influenza surveillance – Sample collection from healthy, sick and dead birds, 2006 (En, Fr, Ru, Ar, Ba, Mn, Es^e, Zh^e, Th)
5. Wild birds and avian influenza – An introduction to applied field research and disease sampling techniques, 2007 (En, Fr, Ru, Ar, Id, Ba)
6. Compensation programs for the sanitary emergence of HPAI-H5N1 in Latin American and the Caribbean, 2008 (En^e, Es^e)
7. The AVE systems of geographic information for the assistance in the epidemiological surveillance of the avian influenza, based on risk, 2009 (En^e, Es^e)
8. Preparation of African swine fever contingency plans, 2009 (En, Fr, Ru, Hy, Ka, Es^e)
9. Good practices for the feed industry – implementing the Codex Alimentarius Code of Practice on good animal feeding, 2009 (En, Zh, Fr, Es, Ar)
10. Epidemiología Participativa – Métodos para la recolección de acciones y datos orientados a la inteligencia epidemiológica, 2011 (Es^e)
11. Good Emergency Management Practice: The essentials – A guide to preparing for animal health emergencies, 2011 (En, Fr, Es, Ar, Ru, Zh, Mn**)
12. Investigating the role of bats in emerging zoonosese – Balancing ecology, conservation and public health interests, 2011 (En)
13. Rearing young ruminants on milk replacers and starter feeds, 2011 (En)
14. Quality assurance for animal feed analysis laboratories, 2011 (En, Fr^e, Ru^e)
15. Conducting national feed assessments, 2012 (En, Fr)
16. Quality assurance for microbiology in feed analysis laboratories, 2013 (En, Zh**)
17. Risk-based disease surveillance – A manual for veterinarians on the design and analysis of surveillance for demonstration of freedom from disease, 2014 (En)
18. Livestock-related interventions during emergencies – The how-to-do-it manual, 2016 (En, Zh**)
19. African Swine Fever: Detection and diagnosis – A manual for veterinarians, 2017 (En, Zh, Ru, Lt, Sr, Sq, Mk, Es)
20. Lumpy skin disease – A field manual for veterinarians, 2017 (En, Ru, Sq, Sr, Tr, Mk, Uk, Ro, Zh)
21. Rift Valley Fever Surveillance, 2018 (En, Fr, Ar)
22. African swine fever in wild boar ecology and biosecurity, 2019 (En, Ru**, Fr**, Es, Zh**, Ko, Lt)
23. Prudent and efficient use of antimicrobials in pigs and poultry, 2019 (En, Ru, Fr**, Es**, Zh**)
24. Good practices for the feed sector - Implementing the Codex Alimentarius Code of Practice on Good Animal Feeding, 2020 (En, Vi**)
25. Good emergency management practice: The essentials – A guide to preparing for animal health emergencies. Third edition, 2021 (En, Es, Ru, Fr, Ar)
26. Guide to conducting After Action Reviews for animal health emergencie, 2022 (En)

Availability: May2022

Ar – Arabic	Ko – Korean	Sr – Serbian	Multil – Multilingual
Ba – Bashkir	Lt – Lithuanian	Th – Thai	* Out of print
En – English	Mk – Macedonian	Tr – Turkish	** In preparation
Es – Spanish	Mn – Mongolian	Uk – Ukrainian	^e E-publication
Fr – French	Pt – Portuguese	Vi – Vietnamese	
Hy – Armenian	Ro – Romanian	Zh – Chinese	
Id – Indonesian	Ru – Russian		
Ka – Georgian	Sq – Albanian		

The *FAO Animal Production and Health Manuals* are available through authorized FAO Sales Agents or directly from Sales and Marketing Group, FAO, Viale delle Terme di Caracalla, 00153 Rome, Italy.

FAO ANIMAL HEALTH MANUALS
1. Manual on the diagnosis of rinderpest, 1996 (En)
2. Manual on bovine spongifom encephalophaty, 1998 (En)
3. Epidemiology, diagnosis and control of helminth parasites of swine, 1998 (En)
4. Epidemiology, diagnosis and control of poultry parasites, 1998 (En)
5. Recognizing peste des petits ruminant – a field manual, 1999 (En, Fr)
6. Manual on the preparation of national animal disease emergency preparedness plans, 1999 (En, Zh)
7. Manual on the preparation of rinderpest contingency plans, 1999 (En)
8. Manual on livestock disease surveillance and information systems, 1999 (En, Zh)
9. Recognizing African swine fever – a field manual, 2000 (En, Fr)
10. Manual on participatory epidemiology – method for the collection of action-oriented epidemiological intelligence, 2000 (En)
11. Manual on the preparation of African swine fever contigency plans, 2001 (En)
12. Manual on procedures for disease eradication by stamping out, 2001 (En)
13. Recognizing contagious bovine pleuropneumonia, 2001 (En, Fr)
14. Preparation of contagious bovine pleuropneumonia contingency plans, 2002 (En, Fr)
15. Preparation of Rift Valley Fever contingency plans, 2002 (En, Fr)
16. Preparation of foot-and-mouth disease contingency plans, 2002 (En)
17. Recognizing Rift Valley Fever, 2003 (En)